Semiconductor Devices

Semiconductor Devices

M. J. Cooke

Prentice Hall
New York   London   Toronto   Sydney   Tokyo   Singapore

First published 1990 by
Prentice Hall International (UK) Ltd
Campus 400, Maylands Avenue
Hemel Hempstead
Hertfordshire, HP2 7EZ
A division of
Simon & Schuster International Group

Typeset in 10/12 pt Times
by Keytec Typesetting Ltd, Bridport, Dorset

Printed and bound in Great Britain by
Redwood Books, Trowbridge, Wiltshire.

Library of Congress Cataloging-in-Publication Data

Cooke, Mike.
  Semiconductor devices / by Mike Cooke.
    p. cm.
  Includes bibliographical references
  ISBN 0-13-806183-1
    1. Semiconductors. I. Title.
  TK7871.85.C619   1990
  621.381'52–dc20   90-34590
                                              CIP

British Library Cataloguing in Publication Data

Cooke, Mike
  Semiconductor devices.
  I. Title
  621.3815
  ISBN 0-13-806183-1

---

  4 5 96 95

Contents

Preface

Another text on semiconductor devices? There were two reasons for writing this book. First, it is a difficult subject for many students. The existing textbooks were comprehensive, for the most part well written, but readily comprehensible only to the top 20 per cent or so of students. In lectures and tutorial classes at University College Swansea it was always necessary to go over the basic ground carefully and repeatedly – and these students had qualifications well above the minimum. This book is an attempt to communicate core concepts to the *majority* of engineering and science students.

Second, it was written to avoid a feeling of waste. The lectures on semiconductor devices given at Swansea were not prepared without effort. An honest attempt was made to get the plus and minus signs right in a self-consistent way, and to think through the behaviour of junctions and devices. (Lecturing is a sure way of probing the limits of your own understanding.) This work seemed, to me at least, worth sharing.

The pattern of the book is conventional, with one major exception: quantum-mechanical considerations are dealt with lightly and qualitatively. Let particle physicists, philosophers and theologians probe the nature and meaning of this subtle universe; engineers have to get things built to specification and within budget. This omission will need to be corrected for those who wish to study and design devices at a graduate level, but let us get through the basics of device behaviour first, without overburdening those whose talents lie elsewhere. More simply, the average electronics syllabus is far too full and something has got to give.

Many thanks to Professor Ken Board, the research students and the undergraduates at Swansea, who will recognize fragments of this book as their own. Felicity and Janice typed Chapters 1–3 and Becky Moras typed the rest. They did not know what they were letting themselves in for but carried the job through with skill and fortitude. Thanks.

1 Fields and forces

1.1 Current density and electric field

Electric current is the flow of electric charge. If a total charge Q coulombs flows evenly past a point in t seconds, the current, I, is:

$$I = Q/t \text{ A} \tag{1.1}$$

More generally, the current can be defined as the rate of charge flow in coulombs/sec and:

$$I = \frac{\mathrm{d}Q}{\mathrm{d}t} \text{ A} \tag{1.2}$$

Thus a current of one ampere is the flow of one coulomb of charge per second. When discussing the current flowing within materials and devices it is more convenient to use the *current density J* (A m^{-2}) to produce formulae which are not tied to a specific size or shape. If a current I flows uniformly in a material of cross-sectional area A m^2, the current density is:

$$J = I/A \text{ A m}^{-2} \tag{1.3}$$

The quantity used in place of voltage is the *electric field*, $\mathscr{E}$ measured in volts per metre. If a voltage, V, is applied between two planes L metres apart, with a uniform uncharged medium in between, the electric field in the medium is:

1

$$\mathscr{E} = V/L \; \mathrm{V\,m^{-1}} \tag{1.4}$$

The medium can be insulating or conductive.

In general if the voltage, V, varies with position x, the electric field is minus the local gradient of potential:

$$\mathscr{E} = -\frac{\mathrm{d}V}{\mathrm{d}x} \; \mathrm{V\,m^{-1}} \tag{1.5}$$

This minus sign means that $\mathscr{E}$ is directed down the potential gradient, pointing from high to low potential. For example, if the voltage increases in the $+x$ direction as in Figure 1.1 $(\mathrm{d}V/\mathrm{d}x)$ is positive and $\mathscr{E}$ is negative.

If the medium is not electrically neutral but carries a net charge of density $\rho(x)$ coulombs $\mathrm{m^{-3}}$, Gauss's law relates this to the electric field. In one dimension the law may be written:

$$\frac{\mathrm{d}\mathscr{E}}{\mathrm{d}x} = \frac{\rho(x)}{\epsilon} \tag{1.6}$$

ϵ is the permittivity and can be split into two parts: ϵ_0, the permittivity of free space, and ϵ_r, the relative permittivity or the dielectric constant for the medium:

$$\epsilon_0 = 8.85 \times 10^{-12} \; \mathrm{F\,m^{-1}} \tag{1.7}$$
$$\epsilon = \epsilon_0\epsilon_r \; \mathrm{F\,m^{-1}} \tag{1.8}$$

Gauss's law (1.6) can be combined with equation (1.5) to link the potential $V(x)$ and the local charge density:

$$\frac{\mathrm{d}^2V}{\mathrm{d}x^2} = \frac{-\rho(x)}{\epsilon} \tag{1.9}$$

This important relationship is known as Poisson's equation. The properties of many semiconductor devices are determined by the variation of potential through charged junction regions, and Poisson's equation must be solved in these regions to determine the I/V characteristics.

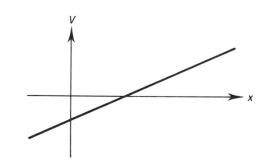

Note:
The line has a positive slope. The electric field is negative and points down the slope.

Figure 1.1 Potential increases with x

Strictly speaking:

1. $\mathscr{E}$ and J are vectors, possessing both magnitude and direction. Restricting the analysis to one dimension enables us to avoid vector notation by using positive quantities to stand for vectors pointing in the $+x$-direction, and negative numbers for vectors pointing in the $-x$-direction.

2. There is another component of current, the *displacement current density*:

$$J_{DIS} = \epsilon \frac{\mathrm{d}\mathscr{E}}{\mathrm{d}t} \qquad (1.10)$$

This only matters when the electric field varies rapidly in time and can be neglected in all device analysis (except for microwave devices).

EXAMPLE

A medium is neutral apart from a constant positive charge density $c \, \mathrm{C m}^{-3}$ in the region $0 < x < d$. The potential at $x = 0$ is zero and there is no electric field at $x = d$. Find and sketch the variation of electric field and potential with x.

☐ First sketch the charge distribution (Figure 1.2).

Use equation (1.6) for the electric field:

$$\frac{\mathrm{d}\mathscr{E}}{\mathrm{d}x} = \frac{\rho(x)}{\epsilon} = \frac{c}{\epsilon}$$

Integrate once with respect to x:

$$\mathscr{E} = \frac{cx}{\epsilon} + C_1$$

C_1 is a constant of integration, which can be found by noting that $\mathscr{E} = 0$ at $x = d$:

$$C_1 = \frac{-cd}{\epsilon}$$

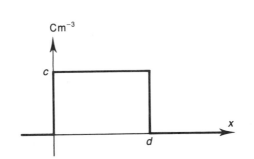

Figure 1.2 Charge distribution

And:

$$\mathscr{E} = \frac{cx}{\epsilon} - \frac{cd}{\epsilon}$$

Equation (1.5) now gives the potential:

$$\mathscr{E} = -\frac{dV}{dx} = \frac{cx}{\epsilon} - \frac{cd}{\epsilon}$$

Integrate again:

$$-V = \frac{cx^2}{2\epsilon} - \frac{cdx}{\epsilon} + C_2$$

The constant of integration C_2 can be found because $V = 0$ at $x = 0$:

$$C_2 = 0$$

Hence:

$$V = \frac{cdx}{\epsilon} - \frac{cx^2}{2\epsilon}$$

So $\mathscr{E}$ and V can be sketched (Figures 1.3 and 1.4).

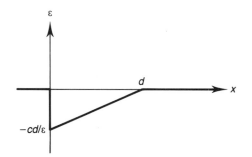

Figure 1.3 Electric field distribution

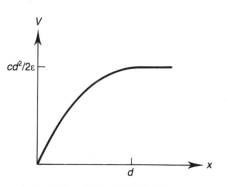

Figure 1.4 Potential distribution

Note that $\mathscr{E}$ is zero and V is constant outside the charged region.

1.2 Forces on charges

In some respects electrons and other fundamental charges behave like very small particles, each with a certain mass and charge. The symbols q and m will be used in this text to signify the magnitude of the charge and mass of an electron at rest in a vacuum:

$$q = + 1.6 \times 10^{-19} \text{ C} \tag{1.11}$$

$$m = 9.11 \times 10^{-31} \text{ kg} \tag{1.12}$$

The charge of an electron is $-q$. An atom which has lost an electron is called a positive ion and has the charge $+q$.

A particle of mass M which experiences a force F newtons will accelerate at $a \text{ m s}^{-2}$ in the direction of the force, as provided by Newton's second law:

$$F = Ma \text{ N} \tag{1.13}$$

In a vacuum all particles move uniformly in straight lines unless acted on by an external force or until they collide. They will obey the usual laws of linear motion relating the distance, s, moved in a time, t, with an initial velocity, u, a final velocity, v, and an acceleration, a, in the direction of motion:

$$v = u + at$$

$$v^2 = u^2 + 2as$$

$$s = ut + \tfrac{1}{2}at^2$$

$$s = \tfrac{1}{2}(u + v)t \tag{1.14}$$

A charge of Q coulombs and mass M in an electric field $\mathscr{E} \text{ V m}^{-1}$ experiences a force:

$$F_E = Q\mathscr{E} \text{ N} \tag{1.15}$$

and hence an acceleration:

$$a = \frac{F_E}{M} = \frac{Q\mathscr{E}}{M} \text{ m s}^{-2} \tag{1.16}$$

A charge Q coulombs moving with a (vector) velocity $\boldsymbol{v}$ in a magnetic field of density $\boldsymbol{B}$ tesla experiences a magnetic field force:

$$\boldsymbol{F}_B = Q\boldsymbol{v} \times \boldsymbol{B} \tag{1.17}$$

$\boldsymbol{F}_B$ is always perpendicular to both $\boldsymbol{B}$ and $\boldsymbol{v}$ because of the vector product, so the particle moves in a circular path, with a *gyro-radius*, R:

$$R = Mv/QB \qquad (1.18)$$

Magnetic field effects are not often encountered in solid state devices, with the exception of Hall effect detectors.

1.3 Energy and the electron-volt

The kinetic energy of a particle of mass M moving with velocity v is:

$$\text{kinetic energy} = \tfrac{1}{2}Mv^2 \text{ J} \qquad (1.19)$$

The potential energy of a charge Q coulombs at a potential V volts is:

$$\text{potential energy} = QV \text{ J} \qquad (1.20)$$

If a charged particle falls freely through a potential drop of V volts, the final velocity is therefore:

$$v = \left(\frac{2QV}{M}\right)^{1/2} \text{ m s}^{-1} \qquad (1.21)$$

Because the charge on an electron is so small its potential energy rarely exceeds 10^{-17} J within a device. A more convenient unit is the *electron-volt* (eV), which is the potential energy of an electron at a potential of one volt. The conversion is:

$$1 \text{ eV} = 1.6 \times 10^{-19} \text{ J} \qquad (1.22)$$

so that energies given in eV must be multipled by q before being used in formulae using SI units.

EXAMPLE

■ An electron is accelerated horizontally from rest through a potential of 200 V. It enters a region of uniform perpendicular electric field between two metal plates 30 mm apart with a potential of 60 V between them. Find the perpendicular deflection and vertical velocity of the electron after travelling a distance x horizontally in the field (Figure 1.5).

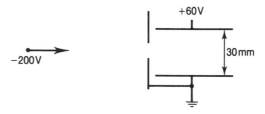

Figure 1.5 Electron acceleration and deflection

☐ The initial horizontal electron velocity u on entering the transverse field is found by equating potential energy loss with kinetic energy gain:

$$\tfrac{1}{2}mu^2 = qV$$

$$u = \left(\frac{2qV}{m}\right)^{1/2} = (2 \times 1.6 \times 10^{-19} \times 200/9.11 \times 10^{-31})^{1/2}$$

$$= 8.38 \times 10^6 \text{ m s}^{-1}$$

Between the plates it is convenient to divide the motion into two components:

1. *Horizontal motion:* there is no further horizontal acceleration, so the horizontal displacement x at time t after going between the plates is:

    $$x = ut$$

2. *Vertical motion:* the electric field between the plates is given by equation (1.4):

    $$\mathcal{E} = \frac{V}{L} = \frac{60}{0.03} = 2{,}000 \text{ V m}^{-1}$$

    The vertical acceleration towards the positive plate is found using equation (1.16):

$$a = \frac{F_E}{m} = \frac{q\mathcal{E}}{m} = 1.6 \times 10^{-19} \times 2{,}000/9.11 \times 10^{-31}$$

$$= 3.51 \times 10^{14} \text{ m s}^{-2}$$

The vertical displacement, y, after a time t is found using $s = ut + \tfrac{1}{2}at^2$:

$$y = 0 + \tfrac{1}{2}at^2$$

The vertical velocity is zero initially. Substitute for t:

$$y = \tfrac{1}{2}a\left(\frac{x}{u}\right)^2$$

$$y = 2.5x^2 \text{ m}$$

The electron traces a parabolic path within the perpendicular field. The vertical component of velocity is given by $v = u + at$:

$$v = 0 + at$$

$$v = \frac{ax}{u}$$

$$v = 4.19 \times 10^7 x \text{ m s}^{-1}$$

Summary

Main symbols introduced

Symbol	Unit	Description
B	T	magnetic flux density
$\mathcal{E}$	V m^{-1}	electric field strength
F	N	force
I	A	electric current
J	A m^{-2}	current density
m	kg	electron mass (9.11×10^{-31})
q	C	size of electron charge (1.6×10^{-19})
Q	C	particle charge
V	V	electric potential
ϵ	F m^{-1}	permittivity of a medium
ρ	C m^{-3}	charge density

Main equations

$$I = \frac{dQ}{dt} \tag{1.2}$$

$$\mathcal{E} = -\frac{dV}{dx} \tag{1.5}$$

$$\frac{d^2V}{dx^2} = \frac{-\rho}{\epsilon} \tag{1.9}$$

$$F_E = Q\mathcal{E} \tag{1.15}$$

Exercises

1. A 13 A current flows in a 2 mm diameter copper wire. Find the current density and the number of electrons passing a point in one second.

2. Find the time taken for an electron to cross between electrode plates 12 mm apart in a vacuum tube, if the potential between them is 90 V. Assume the electron starts at the negative plate with zero velocity.

3. A beam of electrons at zero potential with an initial velocity of 10^6 m s^{-1} is directed

straight at a metal plate in a vacuum. What potential should be applied to the plate to prevent the beam striking it?

4. The potentials at $x = -1$ m and $x = 2$ m are measured to be -400 V and $+200$ V respectively. Find the magnitude and sign of the electric field between these points, assuming it to be constant.

5. The electric field is $30x^2$ V m^{-1} in a region $0 < x < 10$ m, and it is zero elsewhere. Find and sketch expressions for the charge density and the potential, assuming $V = 0$ at $x = 0$. Find the potential at $x = 5$ m. Take $\epsilon = \epsilon_0 = 8.85 \times 10^{-12}$ F m^{-1}

6. The potential V is given by $-1.5 \times 10^{13}x^2$ volts. The charge in the region is due to a uniform distribution of singly charged ions. Find the number of ions per unit volume and their polarity. The dielectric constant is 12.

7. A horizontal beam of electrons travelling at 10^7 m s^{-1} enters a region of vertical electric field of strength 12 kV m^{-1}, due to plates 20 mm apart. Find the total potential between the plates. Find the vertical deflection, the angle of travel of the beam to the horizontal, and the total beam velocity after travelling a distance of 30 mm horizontally within the field.

8. Show that it is justifiable to neglect the gravitational force on an electron in comparison to the force in a field of 1 V m^{-1}. The acceleration due to gravity is 9.81 m s^{-2}.

9. Find the frequency at which the displacement current density is 1 A m^{-2} in a sinusoidally oscillating electric field of peak strength 1 V m^{-1}.

10. Particles of equal charge q and unequal masses M_1 and M_2 are accelerated from rest through a potential V. They enter a region of uniform perpendicular magnetic field B. Will the different particles be deflected by different amounts by the field?

2 The movement of charge

2.1 Materials for electronics

Philosophers make nice distinctions between the properties of materials and the nature of matter; engineers need not. What concerns the engineer is choosing the appropriate combination of materials to achieve a specific task. Civil engineers, for example, select materials on the basis of their movement under mechanical stress. Electronic engineers require material combinations which respond usefully to electrical stress. An applied voltage of V volts is usually regarded as the stress and the response is the electrical current, I amperes.

The goal of this text is to enable the reader to understand how much current flows in response to particular applied voltages for the fundamental electronic materials and devices. This response is often called the *I/V characteristic* of a device.

Although it is sometimes necessary to bear in mind such properties as density, yield stress or brittleness, the electronic engineer is primarily concerned with *electrical conductivity* $(\sigma\,\Omega^{-1}\,\mathrm{m}^{-1})$ or its inverse, *electrical resistivity* $(\rho_e\,\Omega\,\mathrm{m})$:

$$\sigma = \frac{1}{\rho_e}\,\Omega^{-1}\,\mathrm{m}^{-1} \tag{2.1}$$

Note that a high conductivity implies a low resistivity and vice versa. Many substances show a linear increase in current as the applied voltage, V, is increased and their I/V characteristic is a straight line (Figure 2.1). To say that a material has a constant value of resistivity implies this.

Electronic materials can be grouped into three main classes according to their resistivity:

1. *Conductors*: $\rho_e < 10^{-7}\,\Omega\,\mathrm{m}$
2. *Semiconductors*: $10^{-5} < \rho_e < 1\,\Omega\,\mathrm{m}$
3. *Insulators*: $\rho_e > 10^{+8}\,\Omega\,\mathrm{m}$

These limits are only a guide. It is no more possible to set up strict rules on this than it is to divide the population into the rich and the poor. These boundaries do not overlap but it is surprising how few common materials fall in the gaps under normal conditions. All three types of materials have their uses in electronics: conductors define the paths of current flow; insulators prevent current flow; semiconductors in the form of active devices control the sequence and quantity of current.

Notice also the extraordinary range of size of resistivity. The change in resistivity between a metal and a ceramic is at least fifteen orders of magnitude. No other physical property has such a wide variation. For example, the density of the lightest gas and the heaviest element varies only by seven orders of magnitude.

One other property nust be considered: the way in which atoms are arranged within the solid: its *atomic structure*. Again, materials can be grouped into three on the basis of their structure:

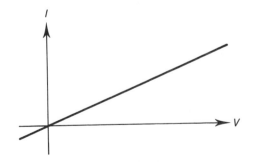

Figure 2.1 The I/V characteristic of a resistive material

1.   *Crystalline*: atoms are arranged in regular patterns.
2.   *Polycrystalline*: the material consists of many small misaligned crystals.
3.   *Amorphous*: atoms are randomly arranged.

It is possible for a material to exist in more than one structure. Silicon, for example, can be deposited in all three forms. Let us be clear at the outset that the active heart of most semiconductor devices requires *crystalline semiconductor* material.

2.2 Unpacking Ohm's law

The remainder of this chapter will deal with the response of non-insulating materials to an applied electric field. The material will be assumed to have uniform properties throughout its volume and will not contain junctions between different sorts of materials.

Consider the case of a uniform rectangular block of such a substance, of cross-sectional area A m^{+2} and length L m, with a voltage V applied across its ends (Figure 2.2). If the resistivity is ρ_e Ωm, the current I which flows is given by:

$$I = \frac{V}{R} = \frac{VA}{\rho_e L} \tag{2.2}$$

where R ohms is the resistance of the block when connected in this way. This equation can also be written:

$$\frac{I}{A} = \frac{1}{\rho_e} \times \frac{V}{L} \; A\,m^{-2} \tag{2.3}$$

or as:

$$J = \sigma\mathscr{E} \; A\,m^{-2} \tag{2.4}$$

J is the *current density* in amperes per square metre, and $\mathscr{E}$ is the *electric field* in volts per metre. This version of Ohm's law is helpful in examining current flows within materials and devices firstly because it is not tied to one

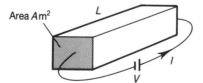

Figure 2.2 Ohm's law gives the resistance of the block as V/I ohms

shape or size of material and secondly because it can be used in situations where the electric field or the conductivity vary through a combination of materials.

2.3 Sign conventions

Strictly, both J and $\mathscr{E}$ are vectors, possessing both magnitude and direction, but as we shall be considering motion along one straight line in most cases the vector notation can be dropped in favour of some sign conventions. These will enable us to work out how much current flows in semiconductor devices in which more than one type of charge is present. The conventions to be used are:

1. *Direction*: positive to the right, the normal graph convention.
2. *Velocity*: positive velocity is motion in a positive direction.
3. *Current*: positive current is given by positive charge moving in a positive direction.
4. *Force*: positive force tends to move a mass in a positive direction.
5. *Electric field*: electric field points from the more positive to the more negative potential, and is positive if it points in a positive direction.

These definitions imply, for example, that an electron travelling to the left is a positive current.

It can be readily shown that the convention for electric field ties up with the normal definition of $\mathscr{E}$ as minus the gradient of electric potential (equation (1.5)). Consider, for example, the potential variation shown in Figure 2.3. The potential falls from 4 V at $x = 1$ to 1 V at $x = 3$. By our rule $\mathscr{E}$ points from $x = 1$ towards $x = 3$, which is the positive direction, and $\mathscr{E}$ is therefore positive. The line on the graph has a negative slope, so equation (1.5) also gives $\mathscr{E}$ as positive.

These conventions have been spelt out in detail because it will be necessary to apply them accurately to get the right ideas about current flows inside semiconductor devices.

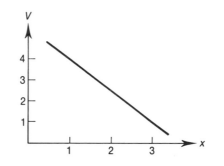

Figure 2.3 Sign convention for electric field

2.4 Drift velocity

Having moved from the circuit parameters of current I and voltage V to the more general variables current density J and electric field $\mathscr{E}$, it is time to look more closely at conductivity, σ. To do this let us consider a tube along which a current of air moves ping-pong balls at a positive drift velocity v_d m s^{-1}. The tube has a cross-sectional area of A m^2 and the balls are uniformly distributed in the tube (Figure 2.4).

We shall now introduce the important concept of *number density*. The number density is the number of particles per unit volume and has the unit m^{-3}. In this case suppose there are N ping-pong balls per unit volume or that the number density of balls is N m^{-3}. The number of balls in one metre of the tube is therefore NA. (Check the units.)

Consider now how many balls pass any point in the tube in one second. Since every ball travels with velocity v_d all balls upstream of the point within a distance of $(v_d \times 1 \text{ second})$ metres will pass the point. Each metre of tube contains NA balls, so the number passing a point in one second is NAv_d. Take this one stage further and suppose that each ball carries a charge of Q coulombs and the charge per second passing a point is $QNAv_d$ coulombs s^{-1} or amperes.

To reiterate:

Number of balls in one metre of tube $\quad = NA$ m^{-1}
Number passing a point in one second $\quad = NAv_d$ s^{-1}
If each ball carries charge Q the current $= QNAv_d$ A

Division by the area A provides us with a second expression for the current density:

$$J = QNv_d \text{ A m}^{-2} \tag{2.5}$$

The velocity v_d in equation (2.5) is called the *drift velocity*. This equation applies in any situation where the number density of mobile charges can be defined and an average velocity can be determined.

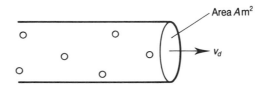

Figure 2.4 N balls m^{-3} each moving at a drift velocity v_d m s^{-1}

2.5 Conductivity and mobility

The two important equations for current density can be compared:

$$J = \sigma\mathscr{E} \; \mathrm{A\,m^{-2}} \tag{2.4}$$

$$J = QNv_d \; \mathrm{A\,m^{-2}} \tag{2.5}$$

It is clear that provided Q, the charge per particle, and N, the particle number density, remain constant there is a relationship between the drift velocity and the electric field of the form:

$$v_d = \mathrm{constant} \times \mathscr{E} \tag{2.6}$$

This constant is called the *mobility* and has units of $\mathrm{m^2\,V^{-1}\,s^{-1}}$ (check it). Mobility is given the symbol μ, sometimes with subscripts to denote which kind of particle is moving. Equation (2.5) can then be written:

$$J = QN\mu\mathscr{E} \; \mathrm{A\,m^{-2}} \tag{2.7}$$

which is the form used in electronic device calculations. A current which flows in response to an electric field is called a *drift current*. A comparison of equations (2.4) and (2.7) shows that the conductivity is:

$$\sigma = QN\mu \; \Omega^{-1}\,\mathrm{m^{-1}} \tag{2.8}$$

and so depends on the number density of mobile charges and their ability to move in response to an applied electric field.

The process of unpacking Ohm's law can be taken one stage further by looking at the mobility. When charged particles are placed in an electric field they experience a force F_E:

$$F_E = Q\mathscr{E} \; \mathrm{N} \tag{2.9}$$

where Q coulombs is the charge per particle as before. Note that our sign convention implies that the force moves a positive charge towards the more negative potential. This force will tend to increase the momentum (mv_d) of the particles, and according to Newton's second law:

$$\mathrm{rate\ of\ increase\ in\ momentum} = Q\mathscr{E} \tag{2.10}$$

Mobile charges in a solid make collisions which impart energy to the solid and remove the excess momentum from the charges. Let us suppose that these collisions occur τ_c seconds apart on average. The average excess momentum of a particle is mv_d, and this is lost every τ_c seconds:

$$\mathrm{rate\ of\ decrease\ in\ momentum} = \frac{mv_d}{\tau_c} \tag{2.11}$$

A balance will occur when the rate of increase equals the rate of loss of momentum, which gives:

$$v_d = \frac{Q\tau_c}{m}\mathscr{E}$$ (2.12)

and hence the mobility can be written:

$$\mu = \frac{Q\tau_c}{m}$$ (2.13)

This analysis gives the correct answer but is not rigorous. An average momentum loss and an average collision interval have been used where the two should have been combined before averaging.

This result means that a negatively charged particle would have a negative mobility. However, virtually every text and reference book lists all mobilities as *positive*, regardless of the particle's polarity. We will follow the majority and rewrite the equation for drift velocity as two equations:

positive charges $v_d = +\mu\mathscr{E}$ m s^{-1}

negative charges $v_d = -\mu\mathscr{E}$ m s^{-1} (2.14)

These equations will be used in the remainder of the book.

2.6 Limits to Ohm's law

If the results of the previous sections are inserted into equation (2.5) we obtain a more detailed version of Ohm's law:

$$J = \frac{NQ^2\tau_c}{m}\mathscr{E}$$ (2.15)

It is now possible to ask to what extent the law is a fundamental truth. For the current density J to rise linearly with the applied electric field $\mathscr{E}$ the following quantities must be constant:

1. *Charge per particle, Q*: in solids the mobile charges nearly always carry a fixed charge equal in magnitude to the charge on an electron.

2. *Particle mass, m*: in conducting solids the electrons which are free to move behave as if they possess an *effective mass m^** slightly different from their mass in a vacuum. This is a quantum-mechanical effect, quite different from the correction to mass in relativity, and is specific to each material. The effective mass arises from averaging the effect of all the electric fields in the vicinity of the charged atomic nuclei. It turns out that all these complex field variations can be accounted simply by altering the mass of the electron. The electron responds to *externally* applied electric fields as if its mass were m^*, and the *internal* fields are ignored. (An analogy can be made with buoyancy – a stone apparently weighs less when submerged in

water: an electron is easier to move in silicon crystals than in a vacuum.) The effective mass is not necessarily constant, particularly in high electric fields in 'multi-valley' semiconductors such as gallium arsenide (GaAs). Figure 2.5 shows the variation in electron drift velocity with applied electric field in GaAs: the fall in drift velocity at high fields points to an increase in the effective mass. (A class of devices called transferred-electron devices make use of this effect.)

3. *Average collision interval*, τ_c: as the drift velocity increases two effects occur. The first is a decrease in the average collision interval because of an increase in the average particle velocity. The second is the appearance of other types of collisions at higher particle velocities, which also reduces the collision interval. This behaviour is dominant in silicon, in which the drift velocity levels off at a *saturation velocity* as the electric field increases, as shown in Figure 2.6.

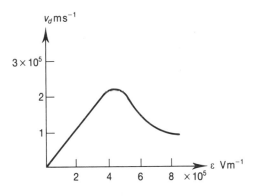

Figure 2.5 The variation of drift velocity with applied electric field in GaAs

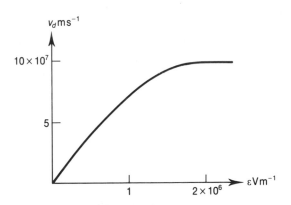

Figure 2.6 Drift velocity variation with electric field in silicon

4.   *Mobile charge number density, N*: at very high fields an electron may gain enough energy between collisions to release another electron from a fixed to a mobile state. This multiplication of charges can escalate rapidly and is called *avalanche breakdown*. The current rises steeply unless limited by a series circuit impedance.

Ohm's law will also fail if current flows for any other reason than an applied electric field. Another mechanism – diffusion – is important in semiconductor devices, as we shall see later.

2.7 Drift and random velocity

It is important to distinguish between drift velocity and *random thermal velocity*, v_{th}, which particles possess by virtue of their thermal energy. A fundamental result of particle kinetic theory is that freely moving particles such as gas molecules have a distribution of speeds. Some move slowly, others quickly, with an average velocity v_{th}. To say that a group of particles has a particular temperature T °K is to say that the speeds are spread out in a *Maxwellian* distribution, in which the number of particles dN with speeds in the small range from v to $(v + dv)$ is given by:

$$dN = \frac{4N}{\sqrt{\pi}} \left(\frac{M}{2kT} \right)^{3/2} v^2 \exp \frac{-Mv^2}{2kT} dv \qquad (2.16)$$

where

N is the total number of particles
M is the particle mass
k is Boltzmann's constant (1.38×10^{-23} J K^{-1})

This imposing expression is sketched in Figure 2.7 for three temperatures ($T_1 < T_2 < T_3$). The vertical axis has the awkward units of (number of particles/velocity interval). It may be easier to look at a thin vertical strip of width dv at any velocity v. The area of the strip is $dv \times (dN/dv)$, which is dN, the number of particles with speeds in this range. The total area under each curve is the same and equals the total number of particles. At higher temperatures there are more particles at higher speeds and the mean speed v_{th} also increases. It can be shown that the mean speed is given by:

$$v_{th} = \left(\frac{8kT}{\pi M} \right)^{1/2} \qquad (2.17)$$

At room temperature the mean thermal speed of an electron is about 10^5 m s^{-1}. This corresponds to a mean thermal energy of 4×10^{-21} joules or 0.026 electron-volts. This last number crops up so often that it is worth learning.

When no electric field is applied mobile charges move randomly in all

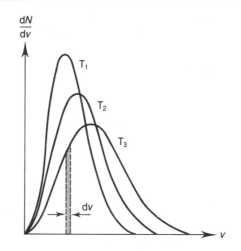

Figure 2.7 The Maxwellian distribution of speeds at three temperatures

directions. There are as many moving to the left as to the right, so there is no net current flow. On the application of an electric field a drift velocity is superimposed on the random velocity. This has two effects on the distribution: it alters its shape and increases the overall mean velocity. If the distance between collisions does not vary it is evident that a constant collision time interval τ_c requires that the drift velocity be small compared to the random velocity: $v_d \ll v_{th}$. If this is not so τ_c falls, and so does the mobility. If the drift velocity approaches the mean thermal velocity, then velocity saturation can be expected, as in Figure 2.6. For this reason a constant value of mobility is sometimes called the *low field mobility*.

2.8 Diffusion

The fact that charged particles may be moving about randomly gives another possible source of current, which does not require the application of an external electric field. Consider the situation in Figure 2.8, in which there are more electrons on the left than on the right. We are assuming for the moment that there are balancing positive charges, so that there are no electric fields present. Random motion will produce a large flow of electrons to the right, which is not countered by an equal flow to the left. This means that there is an overall movement of electrons to the right, which forms a current flow. Currents of this kind are called *diffusion currents*.

In general diffusion will occur in any medium provided:

1. The particles are in random motion.
2. A concentration gradient exists.

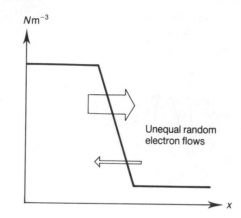

Nm^{-3}

Unequal random
electron flows

x

Figure 2.8 The unequal random fluxes of electrons produce a net diffusion
current at the junction

The diffusion flux F_{diff} – the number of particles crossing a unit area in one
second – is given by Fick's first law:

$$F_{diff} = -D\frac{dN}{dx} \, m^{-2}s^{-1} \tag{2.18}$$

where D is the *diffusion coefficient* for these particles and $N(x)$ is the number
density at position x. The minus sign means that particles diffuse down the
concentration gradient, from high to low number density. In Figure 2.8 no
diffusion occurs in the regions of constant number density, because (dN/dx) is
zero. At the junction (dN/dx) is negative, which gives a positive diffusion flux,
transferring electrons in the $+x$ direction.

If the particles each carry a charge Q the diffusion current which flows is:

$$J_{diff} = -QD\frac{dN}{dx} \, A \, m^{-2} \tag{2.19}$$

which implies, for example, that the diffusion current in Figure 2.8 is negative.
Note that neutral particles can also diffuse. In this case a flux exists, but not a
current.

2.9 The Boltzmann energy distribution

The Maxwellian distribution of kinetic energy is just one way in which energy
can be distributed. Boltzmann showed that the ratio of energy/kT could be used
for potential energy as well. For example, if there are two levels of energy
available with similar properties at E_1 and E_2, the ratio of the populations N_1
and N_2 in the two in equilibrium is:

$$\frac{N_2}{N_1} = \exp\frac{-(E_2 - E_1)}{kT} \tag{2.20}$$

This assumes that the number of particles in each state is not limited. The Boltzmann expression occurs in many other equations, usually in the form:

$$f_B(E) = \exp\frac{-E}{kT} \tag{2.21}$$

where the lower energy level is defined to be zero. It can be thought of as proportional to the fraction of particles with energy E or greater.

The quantity E/kT is dimensionless, being the ratio of energies. It is essential that both energies are expressed in the same units, be they joules (the SI unit), electron-volts or any other. In some texts potential energies are given in (electron-) volts, so the Boltzmann factor used in equations becomes:

$$\exp\frac{-qV}{kT}$$

and the Boltzmann constant is in $J\,K^{-1}$. Worse still, others take the constants k and q together and quote the Boltzmann constant in electron-volts K^{-1}:

$$\frac{k}{q} = \frac{1.38 \times 10^{-23}}{1.6 \times 10^{-19}} = 8.62 \times 10^{-5}\,\text{eV}\,\text{K}^{-1} \tag{2.22}$$

All equations in this text will assume that the energy, E, is in joules and k is in $J\,K^{-1}$. Where problems give energies in electron-volts, these must be multiplied by q to give joules before insertion in the formulae.

EXAMPLE 1

■ Find the maximum resistance of a rectangular block of germanium of dimensions $10\,\text{mm} \times 1\,\text{mm} \times 2\,\text{mm}$ if it can be connected between any pair of parallel faces. Assume there are $10^{21}\,\text{m}^{-3}$ mobile electrons with a charge of $-1.6 \times 10^{-19}\,\text{C}$ each and the electron mobility is $0.39\,\text{m}^2\,\text{V}^{-1}\,\text{s}^{-1}$.

☐ The conductivity is found from equation (2.8):

$$\sigma = QN\mu$$

$$= 1.6 \times 10^{-19} \times 10^{21} \times 0.39$$

$$= 62.4\,\Omega^{-1}\,\text{m}^{-1}$$

Hence the resistivity is:

$$\rho_e = 1/\sigma$$

$$= 0.016\,\Omega\,\text{m}$$

The resistance of the block, using equation (2.2), is:

$$R = \rho_e L / A \ \Omega$$

So for a maximum resistance the largest side is chosen for the length L and the smallest pair for the cross-sectional area A:

$$L = 10 \text{ mm } (0.01 \text{ m})$$

$$A = 1 \text{ mm} \times 2 \text{ mm } (2 \times 10^{-6} \text{ m}^2)$$

The maximum resistance is therefore:

$$R_{max} = 0.016 \times 0.01/2 \times 10^{-6}$$

$$= 80.13 \ \Omega$$

EXAMPLE 2

■ An electron in GaAs has an effective mass of 0.068 m, and makes collisions 0.14 microns apart at 300 K. Find the electron mobility.

☐ The mean thermal velocity v_{th} is given by equation (2.16) and can be found directly:

$$v_{th} = \left(\frac{8kT}{\pi m}\right)^{1/2} = \left(\frac{8 \times 1.38 \times 10^{-23} \times 300}{\pi \times 0.068 \times 9.11 \times 10^{-31}}\right)^{1/2} = 4.125 \times 10^5 \text{ m s}^{-1}$$

The mean time between collisions is found by dividing the mean distance between collisions by the mean velocity:

$$\tau_c = 0.14 \times 10^{-6}/4.125 \times 10^5$$

$$= 3.394 \times 10^{-13} \text{ s}$$

This can be used in the expression for the mobility:

$$\mu = \frac{q\tau_c}{m^*} = \frac{1.6 \times 10^{-19} \times 3.394 \times 10^{-13}}{0.068 \times 9.11 \times 10^{-31}} = 0.877 \text{ m}^2 \text{ V}^{-1} \text{s}^{-1}$$

Note that the sign of the charge is not used because we have defined mobility to be positive.

EXAMPLE 3

■   The potential varies along the x axis of a material, and is measured to be -100 V at the origin, with an electric field of -75 V m^{-1}. Find:

1.   The potential at $x = +0.6$ m.
2.   The direction in which free electrons move.

☐   1.   Two routes can be employed.
  (a)   Use equation (1.6) for the electric field and integrate once with respect to x:

$$\mathscr{E} = -(\mathrm{d}V/\mathrm{d}x) = -75$$

$$V = 75x + C$$

C is the constant of integration, and can be found by putting $V = -100$ at $x = 0$: this gives $c = -100$. Hence at $x = 0.6$ m:

$$V = (75 \times 0.6) - 100$$

$$= -55 \text{ V}$$

  (b)   Use the sign conventions. The electric field is negative, and points in the $-x$ direction:

$$\longleftarrow 75 \text{ V m}^{-1}$$

It points to the more negative potential, so the potential falls at 75 V m^{-1} in the $-x$ direction, or rises at 75 V m^{-1} in the $+x$ direction. In 0.6 m the potential change is:

$$0.6 \times 75 = 45 \text{ V}$$

So in moving from the origin to $x = 0.6$ the potential rises by 45 V:

$$V(0.6) = -100 + 45$$

$$= -55 \text{ V}$$

2.   Electrons will move towards the more positive potential, so they will tend to move from -100 V towards -55 V, in the $+x$ direction.

Summary

Main symbols introduced

Symbol	Unit	Description
D	$m^2\,s^{-1}$	diffusion coefficient
E	J	energy
F_{diff}	$m^{-2}\,s^{-1}$	diffusion flux
k	$J\,K^{-1}$	Boltzmann's constant
		1.38×10^{-23}
m^*	kg	effective mass of a charge in a solid
N	m^{-3}	number density of particles
R	Ω	electrical resistance
T	K	absolute temperature
v_d	$m\,s^{-1}$	average drift velocity
v_{th}	$m\,s^{-1}$	mean thermal velocity
μ	$m^2\,V^{-1}\,s^{-1}$	charged-particle mobility
ρ_e	$\Omega\,m$	electrical resistivity
σ	$\Omega^{-1}\,m^{-1}$	electrical conductivity
τ_c	s	mean time between collisions

Main equations

$$J = \sigma\mathscr{E} \tag{2.4}$$

$$J = QNv_d \tag{2.5}$$

$$\rho_e = 1/QN\mu \tag{2.8}$$

$$\mu = \left(\frac{Q\tau_c}{m}\right) \quad \text{(taking all mobilities as positive)} \tag{2.13}$$

positive charges $v_d = +\mu\mathscr{E}$
negative charges $v_d = -\mu\mathscr{E}$ $\tag{2.14}$

$$J_{diff} = -QD\frac{dN}{dx} \tag{2.19}$$

$$f_B(E) = \exp\frac{-E}{kT} \tag{2.21}$$

Exercises

1. Copper has a resistivity of 16.8 nΩ m. Find the number density of mobile electrons if their effective mass is the free space value, and the mean collision interval is 0.2 ps.

2. Estimate the field at which velocity saturation will occur in gallium phosphide at 300 K if the electron mobility is 0.03 m^2 V^{-1} s^{-1}.

3. Find the drift velocity of electrons in an aluminium cable of diameter 25 mm which carries a current of 2 kA if the resistivity of aluminium is 27 nΩ m and the electron mobility is 0.02 m^2 V^{-1} s^{-1}.

4. A layer of polycrystalline silicon 0.8 microns thick is etched to make a straight track 2 microns wide, which is aligned with the x axis. The potential is measured to be -1 V at the position $x = 2$ microns and -9 V at $x = 26$ microns. Find the electric field, the current density and the current if the electron mobility is 0.1 m^2 V^{-1} s^{-1} and the electron number density is 10^{21} m^{-3}.

5. An electric field of $+80$ V m^{-1} is applied to a solid containing 10^{23} m^{-3} mobile positive charges, each being 1.6×10^{-19} C.
   (a) find the magnitude and direction of the drift velocity and the current density, if the low field mobility is 0.5 m^2 V^{-1} s^{-1}
   (b) find the drift velocity and the current density if:
       (i) the charge polarity is changed;
       (ii) the electric field direction is reversed;
       (iii) both the charge polarity and the electric field direction are reversed.

6. Sketch the I/V characteristics of materials with the following properties:
   (a) the mobility is proportional to the square root of the drift velocity;
   (b) the effective mass is proportional to the drift velocity;
   (c) the collision time interval is inversely proportional to the electric field;
   (d) the mobile charge number density is proportional to the square of the electric field.

7. Show that the conductivity of a material containing N_1 m^{-3} charges carrying $+Q$ coulombs with mobility μ_1 and N_2 m^{-3} charges of $-Q$ coulombs with mobility μ_2 is $Q(N_1\mu_1 + N_2\mu_2)$.

8. Find the diffusion current density if the electron number density varies linearly from 10^{23} m^{-3} to 10^{12} m^{-3} over a distance of 5 microns in the $+x$ direction and the electron diffusion coefficient is 3.5×10^{-3} m^2 s^{-1}.

9. Find the magnitude and sign of the electric field which will produce an equal drift current to the diffusion current in (8) above if the electron mobility is 0.14 m^2 V^{-1} s^{-1}. Use the electron number density at the midpoint of the linear change.

10. Find the fraction of an electron population with an energy of 0.1 eV or more at 300 K.

Hint:

$$\int_0^y x^2 \exp(-x^2)\, dx = \frac{1}{2}\left[\int \exp(-x^2)\, dx - x\exp(-x^2)\right]_0^y$$

and

$$\int_0^y \exp(-x^2)\, dx = y - \frac{y^3}{3} + \frac{y^5}{5 \cdot 2!} - \frac{y^7}{7 \cdot 3!} + \cdots$$

3 | The number of mobile charges

3.1 The atom

The fundamental unit of each of the chemical elements is known as an atom. Atoms of various elements differ greatly in their size and complexity but all share the following features in common:

1. A nucleus, composed principally of neutral particles (neutrons) and particles of charge $+q$ (protons). Neutrons and protons have almost identical masses of about 1.66×10^{-27} kg, nearly 1,700 times the electron mass. Almost all the mass of an atom is concentrated in the nucleus which has a diameter of the order of 10^{-14} m.

2. An electron cloud surrounds the nucleus, extending to a diameter of the order of 10^{-10} m. The number of electrons and protons are equal, so an atom is electrically neutral.

If an atom is placed in an electric field the electrons and protons are pulled in opposite directions. They are held together by the powerful electrostatic attraction between opposite charges, so neither charge can continue to move in response to the field. The small separation of charge which does occur gives rise to the dielectric constant of the material.

If the electrons and protons are attracted to each other, why is it that the electrons in an atom do not all collapse into the nucleus? The answer appears to be that they are constrained by other laws that prevent them taking up the lowest possible positions of energy. These rules of behaviour are known as *quantum mechanics*, in contrast to the classical mechanics of forces and particles reviewed in the first chapter. The main tenets of quantum mechanics are as follows:

1. *The quantization of energy*: electrons can only take certain values of energy, called permitted levels or states, like the rungs of a ladder. An electron has to sit on an allowed rung, not in between the rungs.

2. *The Pauli exclusion principle*: only two electrons are allowed in any one state.

3. Fundamental particles behave in some respects like waves, with a de Broglie wavelength λ:

$$\lambda = \frac{h}{mv}\ \text{m} \tag{3.1}$$

where h is Planck's constant:

$$h = 6.62 \times 10^{-34}\ \text{Js} \tag{3.2}$$

and mv is the momentum of the particle.

4. *Heisenberg's uncertainty principle*: if the position of a particle is known to within Δx and its momentum is known with a precision $\Delta(mv)$, then:

$$\Delta x \Delta(mv) \geq h \tag{3.3}$$

The universe suddenly appears very strange. It is no longer possible to think of measuring the position and momentum of an object with ever increasing precision: the uncertainty principle forbids it. Electrons, like waves, have the ability to 'tunnel' through thin insulating layers which would be insurmountable obstacles if they were only balls of mass and charge. Nevertheless quantum mechanics has gained wide acceptance as a good description of what is actually observed. Theoretical physicists will continue to delve deeper to try to find 'true' laws which describe all observed behaviour precisely. Engineers have to make things work, and must be satisfied with approximations so long as they are good enough: Newton's laws of motion are false because they ignore relativistic effects, but are used by engineers daily because the necessary corrections are insignificant.

Evidence supporting these ideas includes the line spectra derived from atoms given energy by collisions. Atoms shed their excess energy by emitting light at characteristic wavelengths consistent with the light originating in the transition of an electron between discrete levels (Figure 3.1). The same atoms absorb light at these wavelengths more readily, indicating that light energy also comes in packets or photons. The energy E_p of a photon of light of wavelength λ is:

$$E_p = hc/\lambda = hf \text{ J} \tag{3.4}$$

where c is the speed of light $(3 \times 10^8 \text{ m s}^{-1})$ and f is the frequency of the electromagnetic radiation.

The separation of levels on the energy axis corresponds to the distance of an electron from the nucleus, because the potential energy of two charges $+Q$ and $-q$ a distance r apart is given by:

$$\text{potential energy} = \frac{-qQ}{4\pi\epsilon_0 r} \text{ J} \tag{3.5}$$

This is plotted in Figure 3.2, together with the permitted levels of a simple atom. As the charge separation decreases the potential energy falls and the electron becomes ever more tightly bound to the nucleus. The lower levels fill up first and are the last to be emptied by any perturbation. Although each electron would like to get as close as possible to the nucleus, the Pauli exclusion principle forces some to move to higher levels when the lower states are filled.

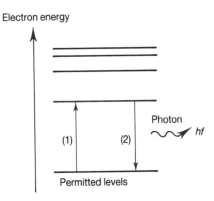

Notes:
(1) Electron gains energy in a collision.
(2) Electron loses energy by emitting a photon.

Figure 3.1 Allowed electron energy states in a single atom

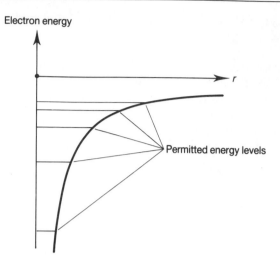

Figure 3.2 Electron energy increases with its separation, r, from the nucleus

This means that some electrons are more loosely bound and are more easily removed from the atom. The outermost electrons are called the *valence* electrons and are responsible for the chemical properties of the atom.

3.2 The band theory of solids

Now consider the situation when two atoms of the kind shown in Figure 3.1 are brought together. The interaction between the electrostatic field of the atoms splits each energy level in two, giving one level slightly higher than before and another a little lower, as shown in Figure 3.3. As more and more atoms are brought together the discrete levels are transformed into a cluster of very narrowly spaced levels or a band of permitted levels of energy. The permitted energy bands are spaced by *band gaps* in which there are no permitted states. The band containing the valence electrons is called the *valence band* and the next permitted band is called the *conduction band*. The top of the conduction band is known as the *vacuum level*. An electron with greater energy than this can escape from the solid completely, as in thermionic emission.

The next point is crucial to the understanding of the band model:

Full bands do not contribute to conduction.

In order to conduct an electron must acquire energy from an applied electric field. If it is to gain energy it must be able to move to a higher energy state. But if all the nearby states are full the electron cannot respond to the field. For this reason we shall not concern ourselves further with the lower energy bands because all conduction takes place in the top two bands.

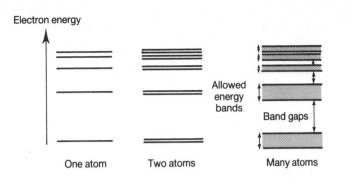

Figure 3.3 The permitted energy levels of electrons form bands of closely spaced levels in a solid

The band model is especially valuable in showing why a material is a conductor, a semiconductor or an insulator. A typical band diagram of the top bands of each type of material is given in Figure 3.4. The top occupied bands in a metal exhibit no band gap, so all the electrons in the top region have empty states close at hand. They can acquire energy from an electric field and hence carry current. This means that the free carrier density is very large. In a semiconductor the top of the valence band is about 10^{-19} J (or 1 eV) below the bottom of the conduction band, and at room temperature a few electrons have sufficient thermal energy to cross the gap. Once in the conduction band they can move freely. A larger band gap of 5×10^{-19} J (~5 eV) or more is characteristic of an insulator. The gap is too large for electrons to cross without external assistance and the number in the conduction band is extremely low.

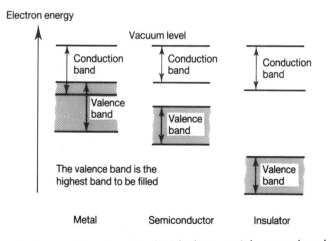

Figure 3.4 The structure of the two top bands in a metal, a semiconductor and an insulator

EXAMPLE

■ Find the maximum wavelength of light which will promote a valence band electron to the conduction band in gallium arsenide, which has a band gap of 1.43 eV.

☐ The minimum photon energy is 1.43 eV or:

$$1.43 \times 1.6 \times 10^{-19} \text{ J} = 2.288 \times 10^{-19} \text{ J}$$

The energy of a photon is given by equation (3.4):

$$E_p = hc/\lambda$$

or:

$$\lambda = hc/E_p$$
$$= 6.62 \times 10^{-34} \times 3 \times 10^8/2.288 \times 10^{-19}$$
$$= 9.68 \times 10^{-7} \text{ m}$$

which is in the near infrared.

3.3 The bond theory

This model looks at the behaviour of valence electrons in solids. Atoms join together to form solids under some conditions because each atom seeks a position of minimum energy. For many substance this is acheived by sharing or exchanging valence electrons in order to give each atom a full quota of valence electrons (frequently *eight*). Elements such as argon which already have a full quota do not join readily with other elements and only form solids at extremely low temperatures. Three broad types of electron sharing are recognised:

1.   *Ionic bonds*: these can occur in solid compounds where one element requires electrons to fill a band and another has a few extra. For example, sodium has one valence electron while chlorine has seven. By transferring one electron across each sodium atom becomes a positive ion and the chlorine atoms become negative ions, each with their outermost bands complete. Electrostatic attraction between the ions keeps them in contact in a regularly spaced array of alternating kinds.

2.   *Covalent bonds*: elemental semiconductors with four valence electrons take the same crystal structure as diamond, in which the atoms are joined by sharing electrons rather than donating them. An ionic bond would require

the transfer of several electrons, so a lower energy configuration occurs when each atom shares a valence electron with its four nearest neighbours, which points to a tetrahedral crystalline structure. Figure 3.5 shows the pattern projected on to a plane surface (the valence electrons are represented by lines).

3. *Metallic bonds*: these are exhibited by elements with a single valence electron, such as copper, sodium, silver and gold. These substances find a minimum energy configuration when they pool their valence electrons. The valence electrons are no longer tied to specific sites, as in ionic or covalent bonding, but are unlocalized and are free to travel throughout the metal. The high ductility and electrical conductivity of the metals are directly related to this form of bonding.

Some substances display a mixture of bond types. Gallium arsenide, for example, is a compound semiconductor formed by the bonding of gallium (valency III) to arsenic (valency V), where the bonds display both an ionic and a covalent character. Several other III–V and II–VI compounds are semiconductors, some of which are listed in Table 3.1 (at the end of this chapter).

3.4 Electrons and holes

The bond model is useful in displaying the origin of the two kinds of mobile charge in a semiconductor: electrons and holes. In a pure silicon crystal at very low temperatures all valence electrons are attached to their home sites and there are no mobile charges (Figure 3.5). At room temperature some electrons have enough thermal energy to break their bonds and become unlocalized free

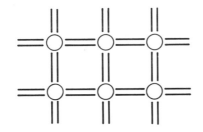

Note:
Each line represents one valence electron shared between two atoms. At low temperature all electrons stay in the bond position.

Figure 3.5 Covalent bonding

electrons (Figure 3.6). This leaves the home site positively charged. It is now possible for other valence electrons to transfer to the vacated sites, without gaining enough energy to break their covalent bonds. This leads to positive ion *sites* (not whole positive ions) free to migrate around the crystal, and to respond to applied electric fields (Figure 3.7). The vacant site is equivalent to a mobile positive charge $+q$, which is called a *hole*. The hole can even be attributed an effective mass m_h and a mobility μ_h.

In terms of the band model a two-tier conduction mechanism exists in a semiconductor. In the upper tier (the conduction band) are those electrons which have escaped from their home sites to become free to move around the crystal. In the lower tier (the valence band) electrons can move only where there are gaps, or we can say that the holes themselves behave like mobile positive charges.

An analogy can be drawn with a two-level road system (Figure 3.8). The upper level is a motorway with plenty of room in the outer lane, and traffic can move freely at high speed. This is equivalent to a sparsely populated conduction band in which all electrons have vacant permitted states nearby. The lower level is more like a city traffic jam, so choked that cars can only move when a gap appears. In this case cars moving to the right can also be seen as gaps moving to the left. This corresponds to hole conduction in a nearly full valence band. It comes as no surprise that electrons are generally more mobile than holes.

This two-tier system of conduction might have remained a laboratory curiosity but for the discovery that the numbers of electrons and holes can be controlled by adding small amounts of certain impurities called dopants to the crystals. The fraction of impurity atoms is extremely small and must be controlled precisely – even a doped semiconductor is typically 99.999% pure. Two types of dopants are used, called *donors* and *acceptors*. Donors add electrons to the conduction band while acceptors add holes to the valence band. Unfortunately it is not possible to do both: if donors and acceptors are both added, they tend to cancel each other out, leaving only the largest impurity as an active dopant. This is still technologically useful, as we shall see, because it means that impurities can be added successively in increasing quantities to create junctions between regions with different doping.

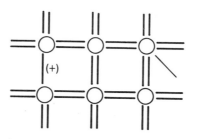

Figure 3.6 Thermal energy can free some valence electrons, leaving a positively charged site (a hole)

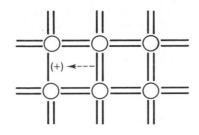

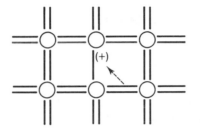

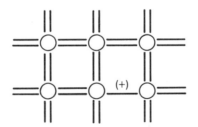

Figure 3.7 Other valence electrons can hop into the vacant site, so the hole can move around the crystal

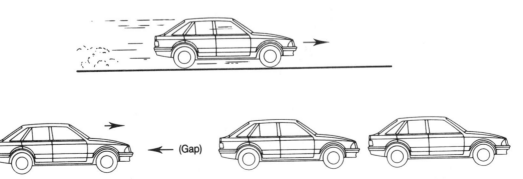

Figure 3.8 A traffic analogy to charge movement in the conduction and valence bands

Any element with *five* valence electrons, such as phosphorus or arsenic, which can take the place of an atom in a semiconductor crystal is a donor impurity. Consider what happens when a donor atom is substituted for a silicon atom (Figure 3.9). Four of its valence electrons are used in covalent bonds with neighbouring atoms but the fifth is redundant. If there is sufficient thermal energy the surplus electron breaks free from the home site and wanders around the crystal. The donor atom was a neutral particle but becomes *ionized* on losing the electron. The material is still electrically neutral because for each positive ion fixed in the lattice there is a negative electron nearby. In terms of the band model, donor impurities add a new donor level close to the conduction band, which was initially full. Thermal energy readily promotes these electrons to the conduction band, where they are *unlocalized* and are free to move about (Figure 3.10).

Similarly an atom with *three* valence electrons which can be substituted for a lattice atom is an acceptor impurity (Figure 3.11). The valence electrons are used in covalent bonds but one site is vacant. This can be filled by a nearby valence electron, creating a fixed negative acceptor ion and a mobile positive ion site – a hole. The initial vacant site can also be regarded as an empty acceptor level close to the valence band, which is easily filled by a valence band electron, leaving a hole (Figure 3.12). If both types of impurities are present we can say that the surplus donor electrons prefer to fill the vacant acceptor sites because they are at a lower energy.

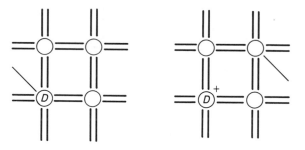

Figure 3.9 Bond model of an impurity donor, *D*

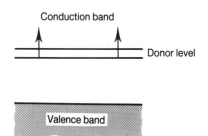

Figure 3.10 Band model of a donor dopant

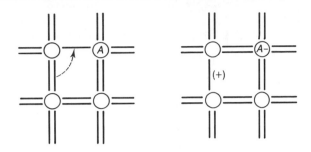

Note:
The neutral atom can acquire an extra electron, creating a mobile hole and a fixed negative ion.

Figure 3.11 Band model for an acceptor impurity

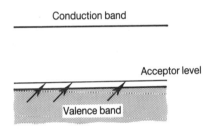

Note:
A new level, initially empty, is added near the full valence band. When it fills it creates holes in the valence band.

Figure 3.12 Band model for an acceptor dopant

3.5 The Fermi–Dirac distribution

The probability that a level at energy E is occupied by an electron is given by the *Fermi–Dirac* distribution $f_{FD}(E)$. It differs from the Boltzmann distribution because it assumes that each level can take only a limited number of particles. It takes the form:

$$f_{FD}(E) = \frac{1}{1 + \exp\dfrac{E - E_F}{kT}} \qquad (3.6)$$

Being a probability $f_{FD}(E)$ has no units. It is plotted against E in Figure 3.13 for different temperatures: nearly all states are full below the Fermi energy E_F and nearly all are empty above E_F. The occupancy of higher energy states increases

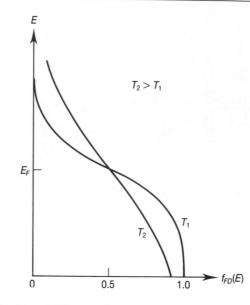

Figure 3.13 The Fermi–Dirac probability of a state at energy E being occupied

with temperature, as would be expected. The probability of a state at the Fermi level being occupied is always one half. The probability of a state being unoccupied is (1 − probability of occupation), which is readily shown to be:

$$1 - f_{FD}(E) = \cfrac{1}{\exp\left(\cfrac{E_F - E}{kT}\right) + 1} \tag{3.7}$$

The actual distribution of particles with energy also depends on the number of states at each energy. The Fermi–Dirac distribution will correspond to the distribution of electrons only if there are a constant number of states in each energy interval E to $E + dE$. This is approximately true in a metal, but is definitely not the case in a semiconductor.

Note that for energies several kT above the Fermi level the exponential term dominates the distribution:

$$\cfrac{1}{1 + \exp\cfrac{(E - E_F)}{kT}} \simeq \exp-\left(\cfrac{E - E_F}{kT}\right) \tag{3.8}$$

so that the Fermi–Dirac distribution tends to the Boltzmann distribution at higher energies, taking E_F as the zero level for energy. The Fermi–Dirac distribution and its Boltzmann approximation will be used to calculate the number of mobile electrons in a semiconductor.

EXAMPLE

■   Find the probability of occupation of a level 0.045 eV above the conduction band edge, if the Fermi level is 0.7 eV above the valence band, the band gap is 1.1 eV and the temperature is 300 K.

☐   The separation between the upper level and the Fermi level in electron-volts is:

$$1.1 - 0.7 + 0.045 = 0.445 \text{ eV}$$

or:

$$E - E_F = 0.445 \times 1.6 \times 10^{-19} \text{ J}$$

$$= 7.12 \times 10^{-20} \text{ J}$$

The probability of this level being occupied is given by the Fermi–Dirac distribution:

$$f_{FD} = (1 + \exp{(E - E_f)/kT})^{-1}$$

$$= 1/(1 + \exp{7.12 \times 10^{-20}/1.38 \times 10^{-23} \times 300})$$

$$= 3.4 \times 10^{-8}$$

This means that nearly all the states in the conduction band at this energy are vacant, and those electrons which are there can move freely.

3.6 The electron number density

In this section a value will be found for the number density of electrons in the conduction band of a semiconductor. A pure semiconductor is called *intrinsic* material, so the subscript *i* will be used to denote its properties. It is clear from the above discussion that for every conduction band electron in an intrinsic semiconductor there is a hole in the valence band:

$$n_i = p_i \text{ m}^{-3} \tag{3.9}$$

where n_i is the intrinsic electron number density and p_i is the intrinsic hole number density. The Fermi level in an intrinsic semiconductor is shown as the intrinsic Fermi level E_{Fi} and is fixed for a given material. A doped semiconductor is known as *extrinsic* material, and the Fermi level E_F will depend on the level and type of doping.

In order to quantify *n* two pieces of information are required:

1.   The number density of states at any energy E.
2.   The probability of any given state being occupied.

The first is given by the *density of states* function. There are so many states at each energy within a permitted band that it is possible to use the methods of calculus and say that there are dN_s states per unit volume in an electron energy range from E to $(E + dE)$. The density of states in the conduction band, N_{sc}, and the valence band, N_{sv}, are approximately given by:

$$dN_{sc} = \left(\frac{4\pi}{h^3}\right)(2m_e^*)^{3/2}(E - E_c)^{1/2}\, dE \ \text{m}^{-3} \tag{3.10}$$

$$dN_{sv} = \left(\frac{4\pi}{h^3}\right)(2m_h^*)^{3/2}(E_v - E)^{1/2}\, dE \ \text{m}^{-3} \tag{3.11}$$

where E_c and E_v are the energies of the conduction and valence band edges. These functions are plotted in Figure 3.14. There are few states at the band edges and an expanding number deeper into the bands. The proof of these formulae is not difficult but involves quantum-mechanical calculations which are not strictly relevant to electronic engineering. (See for example Yang, *Fundamentals of Semiconductor Devices*, Appendix A.)

The probability that a state at an energy E in a solid is occupied is given by the Fermi–Dirac distribution $f_{FD}(E)$, equation (3.6).

A value for the electron number density can now be calculated:

$$n = \int_{\substack{\text{conduction}\\ \text{band}}} \left(\begin{matrix}\text{density of states}\\ \text{at energy } E\end{matrix}\right) \times \left(\begin{matrix}\text{probability}\\ \text{of occupation}\end{matrix}\right) dE$$

$$= \int_{E_c}^{E_{vac}} \left(\frac{4\pi}{h^3}\right)(2m_e^*)^{3/2}(E - E_c)^{1/2} \Big/ \left(1 + \exp\frac{E - E_F}{kT}\right) dE \tag{3.12}$$

This integral is equivalent to the shaded area in Figure 3.15. Although the conduction band does have a top edge (the vacuum level E_{vac}), the rapid fall in the Fermi–Dirac distribution at higher energies means that the integral can take

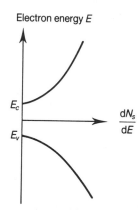

Figure 3.14 Density of states in the conduction and valence bands

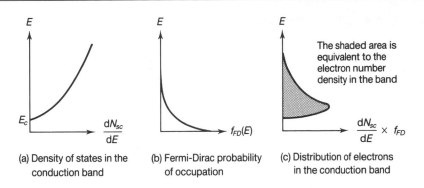

(a) Density of states in the conduction band

(b) Fermi-Dirac probability of occupation

(c) Distribution of electrons in the conduction band

Figure 3.15 States and electrons in the conduction band

an infinite upper limit with negligible error. Assuming the Fermi level is several kT from the conduction band edge, the Fermi–Dirac distribution can be replaced by its Boltzmann approximation, making equation (3.12) soluble analytically:

$$n \simeq \int_{E_c}^{\infty} \left(\frac{4\pi}{h^3}\right)(2m_e^*)^{3/2}(E - E_c)^{1/2} \exp\frac{E_F - E}{kT}\, dE$$

$$n = N_c \exp\frac{E_F - E_c}{kT}\,\mathrm{m}^{-3} \tag{3.13}$$

The coefficients before the exponential have been lumped together to give the *effective density of states* in the conduction band N_c:

$$N_c = \frac{2}{h^3}(2\pi m_e^* kT)^{3/2}\,\mathrm{m}^{-3} \tag{3.14}$$

This is often quoted for particular temperatures, and is typically about $10^{25}\,\mathrm{m}^{-3}$. The remainder of the expression is then equivalent to the fraction of the available states which are occupied. N_c is a weak function of temperature, so most of the variation of n is due to a different fraction of the conduction band states being filled.

A similar analysis can be performed for the intrinsic hole density, using $1 - f_B(E)$ as an approximation to the probability of a state being empty in the valence band. This gives:

$$p = \left(\frac{2}{h^3}\right)(2\pi m_h^* kT)^{3/2} \exp\frac{E_v - E_F}{kT}\,\mathrm{m}^{-3}$$

$$p = N_v \exp\frac{E_v - E_F}{kT} \tag{3.15}$$

where N_v is the effective density of states in the valence band. The expressions for p and n (3.13) and (3.15) cannot be evaluated without knowing the position of the Fermi level with respect to the band edges. However, an important expression can be found by multiplying the equations:

$$pn = N_c N_v \exp \frac{-(E_c - E_v)}{kT} \ \mathrm{m}^{-6} \tag{3.16}$$

This shows that the electron and hole number densities are strongly related to the difference in energy between the valence and conduction bands regardless of the position of the Fermi level. This quantity is called the *band gap*:

$$E_g = E_c - E_v \ \mathrm{J} \tag{3.17}$$

Note that the equations assume E_g is in joules, while it is often quoted in electron-volts. Since p and n are equal in an intrinsic semiconductor, n_i can be found from equations (3.16) and (3.17):

$$n_i = (N_c N_v)^{1/2} \exp \frac{-E_g}{2kT} \ \mathrm{m}^{-3} \tag{3.18}$$

Since the expression resembles the Boltzmann distribution, n_i can be thought of as proportional to the number of electrons with an energy greater than E_g. Clearly a smaller band gap or a higher temperature will lead to an increase in n_i. The position of the intrinsic Fermi level E_{Fi} can be found by equating n and p in equation (3.13) and (3.15):

$$N_c \exp \frac{E_F - E_c}{kT} = N_v \exp \frac{E_v - E_F}{kT} \tag{3.19}$$

Hence:

$$E_{Fi} = \tfrac{1}{2}(E_c + E_v) + \frac{kT}{2} \ln \frac{N_v}{N_c} \tag{3.20}$$

Since the effective densities of states N_c and N_v are nearly equal, E_{Fi} is almost midway between the conduction and valence bands. In general the Fermi level can be found by dividing equation (3.13) and (3.15):

$$E_F = \tfrac{1}{2}(E_c + E_v) + \frac{kT}{2} \left(\ln \frac{N_v}{N_c} + \ln \frac{n}{p} \right) \tag{3.21}$$

or:

$$E_F = E_{Fi} + \frac{kT}{2} \ln \frac{n}{p} \tag{3.22}$$

The Fermi level is thus near the conduction band in n-type material, and near the valence band in p-type material.

3.7 *p*-type and *n*-type semiconductors

Equation (3.16) gives a value for the product of electron and hole number densities in a semiconductor in terms of fixed materials properties and the temperature, and may be written:

$$pn = AT^3 \exp \frac{-E_g}{kT} \tag{3.23}$$

where A is a constant. The righthand side is only a function of temperature and does not depend on the amount of doping. The relation is also true for an intrinsic semiconductor, so we may write:

$$pn = n_i^2 \tag{3.24}$$

where the intrinsic number density at a given temperature is usually a known material property, while p and n are unknown.

An independent condition exists which enables p and n to be calculated: *charge neutrality*. Four sources of charge can exist in a semiconductor:

1. p m^{-3}: mobile valence band holes, charge $+q$.
2. n m^{-3}: mobile conduction band electrons, charge $-q$.
3. N_A m^{-3}: fixed ionized acceptor atoms, charge $-q$.
4. N_D m^{-3}: fixed ionized donor atoms, charge $+q$.

The material is neutral in equilibrium, so the positive and negative charges must balance:

$$p + N_D = n + N_A \tag{3.25}$$

Equations (3.24) and (3.25) should be learnt. If N_A, N_D and n_i are known they can be solved using the quadratic formulae:

$$n = \tfrac{1}{2}[(N_D - N_A) + \sqrt{[(N_A - N_D)^2 + 4n_i^2)]}] \text{ m}^{-3} \tag{3.26}$$

$$p = \tfrac{1}{2}[(N_A - N_D) + \sqrt{[(N_D - N_A)^2 + 4n_i^2)]}] \text{ m}^{-3} \tag{3.27}$$

Equations (3.26) and (3.27) should be used with care, because of the danger of error when subtracting large quantities which are nearly equal.

EXAMPLE

■ Find the electron and hole number densities in a semiconductor if $n_i = 10^{16}$ m^{-3}, $N_D = 10^{20}$ m^{-3} and $N_A = 0$.

☐ Using equation (3.26):

$$n = \tfrac{1}{2}(10^{20} + \sqrt{(10^{40} + 4.10^{32})})$$

$$= \tfrac{1}{2}(10^{20} + 1.00000002 \times 10^{20})$$

$$= 1.00000001 \times 10^{20} \text{ m}^{-3}$$

If equation (3.24) is then used to find p:

$$p = n_i^2/n$$

$$= 10^{32}/1.00000001 \times 10^{20}$$

$$\simeq 1 \times 10^{12} \text{ m}^{-3}$$

Doing the calculation this way, it is not necessary to record so many decimal places. However, if the quadratic formula equation (3.27) is used:

$$p = \tfrac{1}{2}[-10^{20} + \surd(10^{40} + 4.10^{32})]$$

This requires the accurate subtraction of two large numbers which are almost identical. Most pocket calculators will not display the last digit of 1.00000002, and some may even round it off, which would give an incorrect value for p of zero.

Various limiting cases are important:

1. $N_A > N_D$: $N_A - N_D$ is the effective acceptor doping N'_A.
2. $N_D > N_A$: $N_D - N_A$ is the effective donor doping N'_D.
3. $N_D = N_A$: the dopants cancel each other completely and $n = n_i$. This is called a *compensated* semiconductor.
4. $N'_A \gg n_i$: if n_i is neglected in the quadratic formula:

$$p \simeq N'_A; \; n = n_i^2/N'_A \tag{3.28}$$

This is p-type semiconductor material, where most of the mobile charges are positive. Holes are called the *majority carrier* and electrons the *minority* carrier in p-type material.

5. $N'_D \gg n_i$: neglecting n_i again:

$$n = N'_D; \; p = n_i^2/N'_D \tag{3.29}$$

This is n-type semiconductor material, in which most of the current is carried by negative charges. Electrons are the majority carriers and holes the minority carriers in an n-type semiconductor.

If neither of the limits (4) or (5) apply, the full quadratic solution must be used.

π, p^-, p, p^+, p^{++} denote increasing acceptor doping.

ν, n^-, n, n^+, n^{++} denote increasing donor doping.

The terminology of p- and n-type materials refers to the polarity of the majority charge carrier. Let it be emphasized once again that both types of semiconductor are *neutral*: electrons and donor ions are in balance in an n-type medium, holes and acceptor ions in a p-type.

EXAMPLE

■ Find n, p, E_F and E_{Fi} at 300 K if $N_c = 10^{25}$, $N_v = 6 \times 10^{24}$, $E_g = 0.67$ eV, $N_A = 10^{20}$ and $N_D = 8 \times 10^{20}$ m^{-3}. Draw the band diagram.

☐ First use equation (3.18) to find n_i:

$$n_i = (N_c N_v)^{1/2} \exp\frac{-E_g}{2kT} \text{ m}^{-3}$$

$$= (10^{25} \times 6 \times 10^{24})^{1/2} \exp\left(\frac{-(1.6 \times 10^{-19} \times 0.67)}{(2 \times 1.38 \times 10^{23} \times 300)}\right)$$

$$= 1.85 \times 10^{19} \text{ m}^{-3}$$

Note that the band gap in eV is multiplied by q to convert it to joules. The intrinsic Fermi level is found using equation (3.21):

$$E_{Fi} = \tfrac{1}{2}(E_c + E_v) + \frac{kT}{2}\ln\frac{N_v}{N_c}$$

If the valence band edge is taken as the zero for energy ($E_v = 0$), this becomes:

$$E_{Fi} = \tfrac{1}{2}E_g + \frac{kT}{2}\ln\frac{N_v}{N_c}$$

$$= \tfrac{1}{2}(1.6 \times 10^{-19} \times 0.67) + (1.38 \times 10^{-23} \times 300/2) \times \ln(6 \times 10^{24}/10^{25})$$

$$= 5.25 \times 10^{-20} \text{ J or } 0.33 \text{ eV}$$

E_{Fi} is 0.33 eV above the valence band.
To find n and p we must find the effective doping. Since $N_D > N_A$:

$$N_D' = N_D - N_A = 8 \times 10^{20} - 10^{20}$$

$$= 7 \times 10^{20} \text{ m}^{-3}$$

This is more than ten times n_i, which normally satisfies an inequality of the kind $N_D' \gg n_i$, so we can say that the material is *n*-type and:

$$n \simeq N_D' = 7 \times 10^{20} \text{ m}^{-3}$$

$$p = n_i^2/N_D' = 4.89 \times 10^{17} \text{ m}^{-3}$$

The Fermi level can be found using equation (3.22):

$$E_F = E_{Fi} + \frac{kT}{2}\ln\frac{n}{p}$$

$$= 5.25 \times 10^{-20} + (1.38 \times 10^{-23} \times 300/2) \times \ln(7 \times 10^{20}/4.89 \times 10^{17})$$

$$= 6.75 \times 10^{-20} \text{ J or } 0.42 \text{ eV}$$

The band diagram for this material is given in Figure 3.16.

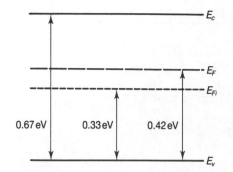

Figure 3.16 Example band diagram

Summary

Main symbols introduced

Symbol	Unit	Description
c	$m\,s^{-1}$	velocity of light in vacuum 3×10^{-8}
E	J	energy
E_c	J	energy at the bottom edge of the conduction band
E_F	J	Fermi energy
E_{Fi}	J	intrinsic Fermi energy
E_g	J	band gap energy (often quoted in eV)
E_p	J	photon energy
E_v	J	energy at the top edge of the valence band
f	Hz	frequency of electromagnetic radiation
h	Js	Planck's constant 6.62×10^{-34}
n	m^{-3}	electron number density in the conduction band
n_i	m^{-3}	intrinsic electron number density
N_c	m^{-3}	effective density of states in the conduction band
N_v	m^{-3}	effective density of states in the valence band
N_s	$m^{-3}\,J^{-1}$	density of states per unit energy
N_{sc}	$m^{-3}\,J^{-1}$	density of states per unit energy in the conduction band
N_{sv}	$m^{-3}\,J^{-1}$	density of states per unit energy in the valence band
p	m^{-3}	hole number density in the valence band
p_i	m^{-3}	intrinsic hole number density
q	C	magnitude of the electron charge 1.6×10^{-19}
r	m	distance between charges
ϵ_0	$F\,m^{-1}$	permittivity of free space 8.85×10^{-12}
λ	m	wavelength of electromagnetic radiation

Main equations

$$E_p = hf \text{ J} \tag{3.4}$$

$$n = N_c \exp \frac{E_F - E_c}{kT} \text{ m}^{-3} \tag{3.13}$$

$$p = N_v \exp \frac{E_v - E_F}{kT} \text{ m}^{-3} \tag{3.15}$$

$$n_i = (N_c N_v)^{1/2} \exp \frac{-E_g}{2kT} \text{ m}^{-3} \tag{3.18}$$

$$E_{Fi} = \tfrac{1}{2}(E_c + E_v) + \frac{kT}{2} \ln \frac{N_v}{N_c} \tag{3.20}$$

$$E_F = E_{Fi} + \frac{kT}{2} \ln \frac{n}{p} \tag{3.22}$$

$$pn = n_i^2 \tag{3.24}$$

$$p + N_D = n + N_A \tag{3.25}$$

Table 3.1 Properties of common semiconductors at 300 K

Quantity		Si	Ge	GaAs	InSb	GaP
Band gap (eV)		1.12	0.67	1.42	0.18	2.26
Effective density of states (m^{-3}):	N_c	2.8×10^{25}	10^{25}	4.7×10^{23}	—	1.7×10^{25}
	N_v	10^{25}	6×10^{24}	7×10^{24}	—	2.2×10^{25}
Intrinsic electron number density, n_i (m^{-3})		1.5×10^{16}	2.5×10^{19}	10^{13}	10^{22}	8×10^6
μ_e electron mobility (m^2 V^{-1}s^{-1})		0.135	0.39	0.85	8	0.03
μ_h hole mobility (m^2 V^{-1} s^{-1})		0.048	0.19	0.045	0.125	0.015
Relative permittivity ϵ_r		11.9	16.0	13.1	17.7	11.1

Exercises

1. The following statements are either true or false:
    (a) An atom which has lost an electron is a positive ion.
    (b) Electrons with the lowest energy stay closest to the atomic nucleus.
    (c) The allowed bands of electron energy all overlap in an insulator.
    (d) More electrons can cross a smaller band gap.

(e)   As the temperature increases fewer electrons can cross the band gap.
(f)   Electronic devices are made of non-crystalline semiconductors.
(g)   The valence band is nearly full of electrons.
(h)   Holes are completely immobile.
(i)   Electrons in the conduction band move like cars in a traffic jam.
(j)   The electron-volt is a unit of charge.
(k)   Longer wavelength photons put electrons in the upper band more easily.
(l)   Completely full bands are unable to carry current.
(m)   High energy states have a high probability of being unoccupied.
(n)   Holes are more likely to be found near the top of the valence band.
(o)   The intrinsic electron number density decreases at higher temperatures.
(p)   The intrinsic hole number density depends on the doping.
(q)   The effective density of states rises with temperature.
(r)   The probability of a hole in a state at energy E is $1 - f_{FD}(E)$.
(s)   The minimum number of electrons in a permitted state is two.
(t)   The intrinsic Fermi level is near the middle of the band gap.

2.   Find the maximum wavelength of light which can create electron–hole pairs in GaP, which has a band gap of 2.26 eV.

3.   Find the equilibrium number densities of electrons and holes in the following semiconductors. The doping densities are zero unless stated otherwise.

	n_i m^{-3}	N_A m^{-3}	N_D m^{-3}
(a) Silicon	1.5×10^{16}	10^{20}	
(b) Germanium	2.5×10^{19}		10^{22}
(c) Gallium arsenide	10^{13}	10^{15}	2×10^{15}
(d) Germanium	2.5×10^{19}	2.5×10^{19}	2.5×10^{19}
(e) Silicon	1.5×10^{16}	10^{20}	10^{19}
(f) Gallium arsenide	10^{13}		5×10^{12}

4.   For each case in the last question find the position of the intrinsic Fermi level and the Fermi level with respect to the edge of the valence band. Take $T = 300$ K, and use the effective densities of states given in Table 3.1.

5.   The band gap in silicon is 1.12 eV, and n_i is 1.5×10^{16} m^{-3} at 300 K. At what temperature does n_i become 10^{20} m^{-3}, assuming:
(a) that N_c and N_v do not vary with temperature;
(b) that they do vary.
(This puts a limit on the maximum temperature at which a silicon device can be used, if the minimum doping is 10^{20} m^{-3}.)

6.   Rank the materials in (3) in order of increasing resistivity, using the electron and hole mobility data in Table 3.1.

4 | Semiconductors in equilibrium

4.1 The combined current equations

In Chapter 2 we saw how current could flow in response to either a concentration gradient of charged particles or an electric field (a potential gradient). These two flows were called diffusion currents and drift currents respectively. The diffusion current density of particles with a number density N and a charge per particle of Q is:

$$J_{diff} = -QD\frac{dN}{dx} \, \text{A m}^{-2} \tag{4.1}$$

The drift current depends on the mean drift velocity:

$$J_{drift} = QNv_d \, \text{A m}^{-2} \tag{4.2}$$

and the drift velocity is proportional to the electric field:

$$v_d = +\mu\mathscr{E} \quad \text{(positive charges)}$$
$$v_d = -\mu\mathscr{E} \quad \text{(negative charges)} \tag{4.3}$$

taking the mobilities as positive. Both polarities of mobile charge exist in a semiconductor: there are n electrons m^{-3} each with a charge $-q$, and p holes m^{-3}, charged $+q$. Electrons and holes have different mobilities, μ_e and μ_h, and different diffusion coefficients D_e and D_h respectively. The electron and hole current densities can each have a drift component and a diffusion component:

$$J_e = qn\mu_e \mathscr{E} + qD_e\frac{dn}{dx} \,\mathrm{A\,m^{-2}} \tag{4.4}$$

$$J_h = qp\mu_h \mathscr{E} - qD_h\frac{dp}{dx} \,\mathrm{A\,m^{-2}} \tag{4.5}$$

The total current is the sum of the two:

$$J = J_e + J_h \,\mathrm{A\,m^{-2}} \tag{4.6}$$

Note that in a uniform semiconductor the concentration gradient terms are zero, and only a drift current flows:

$$J = q(n\mu_e + p\mu_h)\mathscr{E} \,\mathrm{A\,m^{-2}} \tag{4.7}$$

So the conductivity of a uniform semiconductor is:

$$\sigma = q(n\mu_e + p\mu_h) \,\Omega^{-1}\,m^{-1} \tag{4.8}$$

EXAMPLE

■ Find the resistance of a silicon resistor, 100 μm long and 5 μm^2 in cross-section, if it is uniformly doped n-type with $N_D = 10^{20}$ m^{-3}, $n_i = 10^{16}$ m^{-3}, $\mu_e = 0.135$ m^2 V^{-1} s^{-1} and $\mu_h = 0.05$ m^2 V^{-1}s^{-1}.

☐ The material is doped uniformly, so there is no diffusion current. Since $N_D \gg n_i$, we may approximate:

$$n \simeq N_D = 10^{20} \text{ m}^{-3}$$
$$p \simeq n_i^2/N_D = 10^{12} \text{ m}^{-3}$$

The conductivity is given by equation (4.8):

$$\sigma = 1.6 \times 10^{-19}(10^{20} \times 0.135 + 10^{12} \times 0.05)$$
$$= 2.16 \ \Omega^{-1}\,m^{-1}$$

Note that hole conduction in this n-type material is negligible. Most of the current is transported by the majority carrier, electrons in this case. The resistance is therefore:

$$R = \frac{\rho_e L}{A} = \frac{L}{\sigma A} = \frac{100 \times 10^{-6}}{2.16 \times 5 \times 10^{-12}}$$
$$= 9.26 \ \mathrm{M\Omega}$$

4.2 The Einstein relation

A useful relation between the mobility and the diffusion coefficient of a particle can be deduced from the combined current equations by expressing both the electric field and the particle number density in terms of the local potential. To do this we recall two results from the previous chapter:

$$E_F = E_{Fi} + \frac{kT}{2} \ln \frac{n}{p} \tag{4.9}$$

$$pn = n_i^2 \tag{4.10}$$

Substituting first for n and then for p between these equations we get an alternative pair of equations for n and p:

$$n = n_i \exp \frac{E_F - E_{Fi}}{kT} \tag{4.11}$$

$$p = n_i \exp \frac{E_{Fi} - E_F}{kT} \tag{4.12}$$

These equations mean that the electron and hole number densities are raised or lowered from their intrinsic value according to the position of the Fermi level E_F with respect to E_{Fi}. If E_F is above E_{Fi} then n is increased and p is reduced, making the material n-type. In p-type material E_F is below E_{Fi}.

To express n and p in terms of potential, we need to convert equations (4.11) and (4.12) from energies in joules to potentials in volts. Accordingly let us define two potentials with respect to some constant arbitrary zero level:

$$-qV = E_{Fi} \tag{4.13}$$

$$-q\phi = E_F \tag{4.14}$$

V is the local electrostatic potential, which we have identified with E_{Fi}. It could just as well have been defined with some other level, such as the vacuum level, provided the level were fixed relative to the band edges. ϕ is the Fermi potential, which might also vary with position. The minus signs mean that the scales of electron energy and voltage are reversed: the highest electron energies are at the most negative potentials, because the electron carries a negative charge. Figure 4.1 shows the equivalence between the energy and voltage scales for an n-type semiconductor.

Using equations (4.13) and (4.14), equations (4.11) and (4.12) become:

$$n = n_i \exp \frac{q(V - \phi)}{kT} \tag{4.15}$$

$$p = n_i \exp \frac{q(\phi - V)}{kT} \tag{4.16}$$

These relations provide the link between the mobile charge number densities

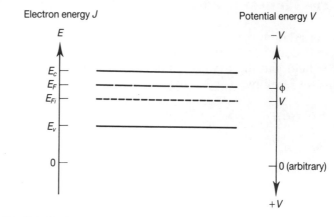

Note. $E_c, E_F, E_{Fi}, E_v > 0$; ϕ, $V < 0$

Figure 4.1 Energies and potentials in an *n*-type semiconductor

and the potential. To find the concentration gradient of electrons, we differentiate equation (4.15) with respect to x:

$$\frac{\mathrm{d}n}{\mathrm{d}x} = \frac{nq}{kT}\left(\frac{\mathrm{d}V}{\mathrm{d}x} - \frac{\mathrm{d}\phi}{\mathrm{d}x}\right) \tag{4.17}$$

where (4.15) has been used again to simplify the expression. Inserting this and equation (1.5) for the electric field into equation (4.4) gives:

$$J_e = -nq\mu_e\frac{\mathrm{d}V}{\mathrm{d}x} + \frac{q^2 nDe}{kT}\left(\frac{\mathrm{d}V}{\mathrm{d}x} - \frac{\mathrm{d}\phi}{\mathrm{d}x}\right) \tag{4.18}$$

There are two ways in which this current can be made zero:

1. No spatial variation of either V or ϕ. The material is uniform and there is no applied electric field, so neither a drift current nor a diffusion current exists.

2. A balance between drift and diffusion currents. Since V and ϕ are independent, at zero current equation (4.18) can be split into two conditions:

$$\frac{\mathrm{d}\phi}{\mathrm{d}x} = 0 \tag{4.19}$$

$$\left(-nq\mu_e + \frac{q^2 nD_e}{kT}\right)\frac{\mathrm{d}V}{\mathrm{d}x} = 0 \tag{4.20}$$

Equation (4.19) can be integrated, using equation (4.14), to give:

$$\phi = \text{constant} = -E_F/q \tag{4.21}$$

which is an important result. It means

The Fermi level is constant in equilibrium.

This principle will be used extensively when drawing the band diagrams of devices.

Equation (4.20) can be simplified, because dV/dx cannot be zero if a drift current exists. It becomes:

$$\frac{D_e}{\mu_e} = \frac{kT}{q} \tag{4.22}$$

This is the Einstein relation and should be memorized. It is left as an exercise to show, using equations (4.5) and (4.16), that an identical relation exists for holes.

EXAMPLE

■   The electron number density in a semiconductor varies from 10^{20} m^{-3} to 10^{12} m^{-3} linearly over a distance of 4 μm. Find the electron diffusion current and the electric field at the midpoint if no current flows, $\mu_e = 0.135$ m^2 V^{-1} s^{-1}, and $T = 300$ K.

☐   The diffusion current of electrons, using the Einstein relation is:

$$J_{ediff} = qD_e\frac{dn}{dx} = \mu_e kT\frac{dn}{dx}$$

Therefore:

$$J_{ediff} = 0.135 \times 1.38 \times 10^{-23} \times 300 \times \frac{10^{20} - 10^{12}}{4 \times 10^{-6}}$$

$$= 1.40 \times 10^4 \text{ A m}^{-2}$$

Using equation (4.4), if J_e is zero:

$$\mathscr{E} = \frac{1}{qn\mu_e}J_{ediff}$$

At the midpoint of this region $n = 0.5 \times 10^{20}$ m^{-3}, so:

$$\mathscr{E} = 1.4 \times 10^4/(1.6 \times 10^{-19} \times 5 \times 10^{19} \times 0.135)$$

$$= 1.29 \times 10^4 \text{ V m}^{-1}$$

4.3 Equilibrium band diagrams

The band diagram is as important to the understanding of solid state devices as the circuit diagram is to electronics. It is not an end in itself because very few engineers will ever be asked to design a band diagram, but it is the single most important tool for the analysis of junctions. So far we have studied the band diagram of a uniform semiconductor such as the one in Figure 4.1, which shows the electron energy (E or $-qV$) as a function of position. It includes the following:

1. *The conduction band*: normally nearly vacant, it contains n mobile electrons m^{-3}. It is bounded by the vacuum level, E_{vac}, and the lower energy limit, E_c.

2. *The band gap*: there are no permitted levels here in a perfect semiconductor. Doping levels can be created in the gap by impurities, but these are not usually shown. The intrinsic Fermi level E_{Fi} and the Fermi level E_F are normally within the gap. These are mathematical quantities, and do not signify the presence of a permitted state.

3. *The valence band*: this is full, but for p holes m^{-3}, which behave like mobile positive charges. Only the top edge of the valence band E_v need be considered.

Electrons and holes have opposite charges, so while the electron energy increases up the diagram, the hole energy increases at the bottom. The following principle may be obvious, but it needs to be spelt out:

Particles tend to take up positions of minimum energy.

Thus if there is no thermal energy, all the electrons are in the valence band. At room temperature, as we have seen, a small fraction of electrons have enough energy to be promoted to the conduction band. There they tend to take up states within a few kT of the lower band edge. The holes they leave behind will also tend to minimize their energy, and cluster near the upper edge of the valence band. Figure 4.2 illustrates this principle: electrons behave like little balls, always rolling down the hill to the lowest position; holes are like bubbles in the valence band, rising to the highest levels possible.

The second principle governing the construction of band diagrams is the constancy of the Fermi level in equilibrium. This was demonstrated mathematically in the last section, but can also be supported by an intuitive argument.

Consider the case in Figure 4.3, in which we have assumed that the Fermi level is *not* constant. An electron will have the same probability, according to the Fermi–Dirac distribution, of occupying the levels A and B, because both are the same height E_1 above the Fermi level. An electron at A will therefore move

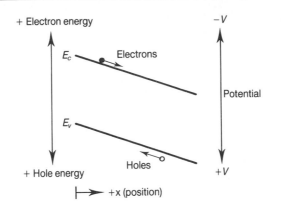

Note. Electrons seek more positive potentials, holes seek more negative potentials.

Figure 4.2 The behaviour of charges in a band diagram

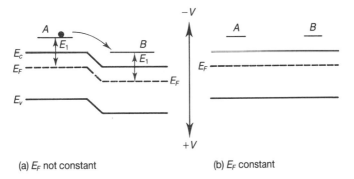

(a) E_F not constant   (b) E_F constant

Figure 4.3 Electrons transfer from A to B, B charges negatively until the Fermi levels align

to B as given by the first principle. This causes the semiconductor at B to become more negatively charged, which raises the energy of *all* the levels at B by the same amount. (A negatively charged region acquires a negative potential, which raises the energy of all negative charges in the region.) Electrons will continue to flow from A to B until the probability of being in each level is the same, which occurs when the Fermi level is the same everywhere. Thus at equilibrium, when no current is flowing, the Fermi level is constant. We now derive the band diagrams for several common situations.

4.3.1 Non-uniform doping

If the distribution of doping impurities in a semiconductor is non-uniform some regions will have a higher concentration of mobile charges than others. This leads to a diffusion current starting to flow but it does not continue for long. Each region is initially neutral, the mobile charges being balanced by the fixed impurity ions. When some of the mobile charges diffuse away, the ions are left partly uncompensated, and the region receiving the mobile charges also acquires a charge. For example, Figure 4.4 shows the distribution of charges before and after diffusion in a non-uniform n-type semiconductor. When a net charge exists in a region, Gauss's law (equation (1.6)) shows that an electric field is set up. This built in field opposes the further diffusion of charges and in equilibrium the drift and diffusion currents will sum to zero.

If we can assume that only a small fraction of mobile charges diffuse away to cause the balancing electric field, equations (4.11) and (4.12) give the local carrier densities. The value of E_{Fi} at any position is thus given by either of these expressions:

$$E_{Fi} = E_F - kT \ln \frac{n}{n_i} \tag{4.23}$$

$$E_{Fi} = E_F + kT \ln \frac{p}{n_i} \tag{4.24}$$

The band diagram can be constructed as follows:

1.   Draw a constant (horizontal) line for the Fermi level E_F.

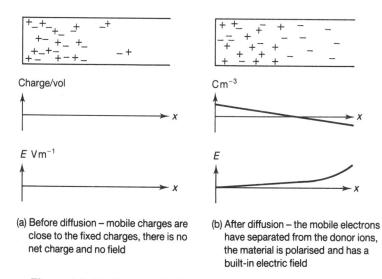

(a) Before diffusion – mobile charges are close to the fixed charges, there is no net charge and no field

(b) After diffusion – the mobile electrons have separated from the donor ions, the material is polarised and has a built-in electric field

Figure 4.4 Distribution of charges before and after diffusion

2. Find n or p as a function of position.
3. Use equation (4.23) or (4.24) to draw E_{Fi} relative to E_F.
4. Draw the band edge energies E_c and E_v parallel to E_{Fi}, with E_{Fi} about halfway in between (equation (3.20)).

EXAMPLE

■ Draw the band diagram of a non-uniform n-type semiconductor in equilibrium, if the doping $N_D = N_0 \exp(-x/L)$. $N_0 = 10^{25}$ m^{-3}, $L = 1$ μm, $T = 300$ K, $n_i = 10^{16}$ m^{-3} and the band gap is 1.1 eV. Find the value of the built in field at $x = 0$.

□ Assume $n \simeq N_D$. This will only be true provided $N_D \gg n_i$. Equation (4.23) then gives:

$$E_{Fi} = E_F - kT \ln \frac{N_0 \exp(-x/L)}{n_i}$$

Therefore:

$$E_{Fi} - E_F = kT\left(\frac{x}{L} - \ln\frac{N_0}{n_i}\right)$$

$$= 4.14 \times 10^{-21}(10^6 x - 20.7) \text{ J}$$

or

$$= 0.026(10^6 x - 20.7) \text{ eV} \tag{4.25}$$

E_{Fi} starts below E_F and rises linearly towards it. The lines will not cross, because the approximation $n \simeq N_D$ will fail before then. To construct the band diagram:

1. Draw a constant Fermi level (Figure 4.5a).
2. Draw E_{Fi}, according to equation (4.25). Figure 4.5b).
3. Draw E_c and E_v parallel to E_{Fi} (Figure 4.5c).

The built in electric field is constant, because E_{Fi} has a constant slope in this case. The local potential V, from equation (4.25) is:

$$V = -2.6 \times 10^4 x \text{ V}$$

where E_F is used as the zero level. The electric field is therefore:

$$\mathscr{E} = \frac{-dV}{dx} = 2.6 \times 10^4 \text{ V m}^{-1}$$

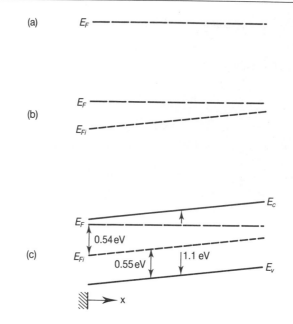

Figure 4.5 Band diagram of non-uniform doping: $N_D = N_0{}^{-x/L}$

It is useful to learn to read band diagrams such as Figure 4.5c. The Fermi level is constant so the material is in equilibrium and there are no applied voltages. The Fermi level is closer to the conduction band towards the left, so the material is more heavily n-type there. The slope on E_{Fi} and E_c indicates a built in field, which tends to confine electrons to the left and prevent diffusion.

4.3.2 Applied electric field

If a constant electric field is applied to a uniform material a drift current will flow. The material is not in equilibrium so E_F is no longer constant. All the equations derived for n and p are only strictly true in equilibrium, but they can continue to be used provided n and p are little changed. If n is still constant through the material, differentiating equation (4.11), we find:

$$\frac{n_i}{kT}\left(\frac{\mathrm{d}E_F}{\mathrm{d}x} - \frac{\mathrm{d}E_{Fi}}{\mathrm{d}x}\right)\exp\frac{E_F - E_{Fi}}{kT} = 0 \tag{4.26}$$

or

$$\frac{dE_F}{dx} = \frac{dE_{Fi}}{dx} \tag{4.27}$$

That is, the Fermi level and the intrinsic Fermi level have the same slope, so remain parallel. Figure 4.6 shows a uniform p-type semiconductor with an applied field, together with the direction of motion of holes and electrons.

4.3.3 Charged regions

Suppose all the mobile charges are removed from a small region of an intrinsic semiconductor. The fixed impurity ions will be left, and the region will have a net charge, positive for donor ions and negative for acceptors. This occurs at junctions between oppositely doped semiconductors and is called a *depletion region* because the zone is depleted of mobile charges. Accumulation regions can also exist, in which the number density of the majority carrier exceeds its equilibrium value. Depletion regions are easier to analyse because the charge distribution is the same as the distribution of dopants.

Figure 4.7a shows the charge distribution in a depletion region in a uniformly doped n-type semiconductor. There are $qN_D \, \text{C m}^{-3}$ in the region $0 < x < d$. There is often a field-free region at one edge of a depletion region, so we shall assume $dV/dx = 0$ at $x = d$. The potential distribution is found by solving Poisson's equation, (1.9):

$$\frac{d^2V}{dx^2} = \frac{-qN_D}{\epsilon} \quad 0 < x < d \tag{4.28}$$

Therefore:

$$\frac{dV}{dx} = \frac{-qN_D x}{\epsilon} + C_1 \tag{4.29}$$

The boundary condition at $x = d$ makes $C_1 = qN_D d/\epsilon$. Integrating again:

$$V = \frac{-qN_D x^2}{2\epsilon} + \frac{qN_D xd}{\epsilon} + C_2 \tag{4.30}$$

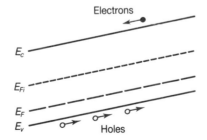

Figure 4.6 A uniform p-type semiconductor with an applied field, causing a drift current to flow

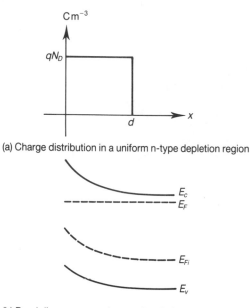

(a) Charge distribution in a uniform n-type depletion region

(b) Band diagram, assuming no electric field for $x > d$

Figure 4.7 Charge distribution and band diagram for a uniform n-type depletion region

The zero point of potential is arbitrary, so if $V(0) = 0$, $C_2 = 0$. Hence E_{Fi} is known:

$$E_{Fi} = -qV = \frac{q^2 N_D}{2\epsilon}(x^2 - 2xd) \tag{4.31}$$

The band diagram is constructed as before (Figure 4.7b):

1. Draw a constant Fermi level, if the device is in equilibrium
2. Draw the variation of E_{Fi} around it. $E_F - E_{Fi}$ outside the charged zone is obtained from equation (4.23).
3. Draw E_c and E_v either side of E_{Fi}.

A constant value of E_{Fi} indicates uniform potential, a constant slope indicates a neutral region with a constant electric field and a curved line shows that the region is charged. It is easy to show that positive charge gives a downwards curvature to the band diagram, while a negative charge gives an upwards curvature. The greater the charge density, the tighter the curve. Figure 4.8 suggests a memory aid for this.

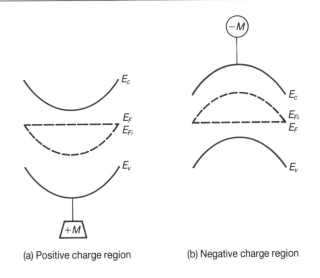

(a) Positive charge region   (b) Negative charge region

Figure 4.8 Positive and negative charged regions produce band diagrams with opposite curvature

4.3.4 Junctions

The analysis of different kinds of junctions will occupy a good deal of an introductory course in semiconductor devices. Although each junction has its own unique properties, the same basic rules apply to the construction of their band diagrams. Three steps are involved:

1. Draw the band diagrams of the two sides of the junction as if the materials were out of contact, aligning the vacuum levels. An electron just extracted, for example by thermionic emission, from either material would be indistinguishable and possess the same energy. Therefore the vacuum energy levels of the two materials are the same before a junction is formed.

2. Start a fresh diagram by drawing a constant Fermi energy level, and add the other energy levels of the materials above and below this, with the two materials side by side. This depicts the shift in levels which occurs upon contact. The Fermi level is constant in equilibrium, not only in uniform or non-uniform materials, but across junctions of dissimilar materials. This is achieved physically by mobile charges moving across the junction when contact is made. They move by diffusion, if concentration gradients exist, and they move to minimize their potential energies. The movement of charges causes regions which were neutral before contact to acquire a

charge, which alters the potential energy of every level in that material, including E_F, until E_F is the same on either side. A balance then exists between drift and diffusion currents, halting further charging.

3. Draw curved lines in the region of the junction to depict the charges which have formed there. Where possible, join up similar bands (conduction or valence bands) with smooth curves.

EXAMPLE

■ Draw the band diagram of a pn^+ junction in equilibrium.

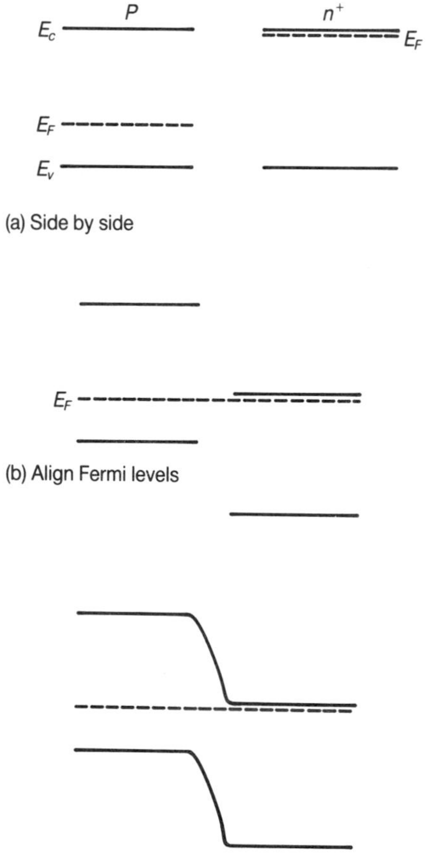

(a) Side by side

(b) Align Fermi levels

(c) Draw curved parts corresponding to charges

Figure 4.9 Drawing the band diagram of a junction

□ 1.   Draw the band diagram for p-type and heavily doped n-type semiconductor side by side, with the vacuum levels aligned (Figure 4.9a). The Fermi level is closer to the band edge in the n^+ material.

2.   Draw the same levels around a constant Fermi level (Figure 4.9b).

3.   Join the bands with smooth curves (Figure 4.9c). Note that the n^+ region has more charges/vol and therefore has a tighter curvature than the p region. The charged regions stay close to the metallurgical junction because the electrostatic attraction between charges prevents them spreading out. The n^+ region has acquired a positive charge, because electrons have diffused out and holes have diffused in. This means that the potential energies of all electrons in the n^+ region have been depressed, causing the Fermi levels to align

This procedure will be explained in greater detail in Chapter 6 for the pn junction, Chapter 9 for metal–semiconductor junctions, and in Chapter 14 for metal–insulator–semiconductor multiple junctions.

Summary

Main symbols introduced

Symbol	Unit	Description
$V(x)$	V	potential at position x (intrinsic Fermi potential)
$\phi(x)$	V	Fermi level potential at position x

Main equations

$$J_e = qn\mu_e\mathscr{E} + qD_e\frac{dn}{dx}\ \mathrm{A\,m^{-2}} \tag{4.4}$$

$$J_h = qp\mu_h\mathscr{E} - qD_h\frac{dp}{dx}\ \mathrm{A\,m^{-2}} \tag{4.5}$$

$$\sigma = q(n\mu_e + p\mu_h)\ \Omega^{-1}\,\mathrm{m^{-1}} \tag{4.8}$$

$$n = n_i\exp\frac{E_F - E_{Fi}}{kT} = n_i\exp\frac{q(V - \phi)}{kT} \tag{4.11 and 4.15}$$

$$p = n_i \exp \frac{E_{Fi} - E_F}{kT} = n_i \exp \frac{q(\phi - V)}{kT}$$ (4.12 and 4.16)

$E_F = $ constant in equilibrium   (4.21)

$$\frac{D_e}{\mu_e} = \frac{kT}{q} \quad \text{(the Einstein relation)}$$ (4.22)

Exercises

1. Find the resistivity of silicon doped with 10^{19} m^{-3} donors and 10^{23} m^{-3} acceptors. Take $\mu_e = 0.135$ and $\mu_h = 0.048$ m^2 V^{-1} s^{-1}.

2. Find the ratio of electron drift and diffusion currents in GaAs at 300 K if the electric field is 10^5 V m^{-1}, the concentration gradient is 10^{26} m^{-4} and the electron number density is 10^{20} m^{-3}.

3. Show, using equations (4.5) and (4.16), that the Einstein relation (4.22) also applies to holes.

4. Sketch the band diagram of the p-type region with non-uniform doping defined by $N_A(x) = 10^{24} \exp -(x/L)$, where $L = 10$ microns. Find the separation in eV between the Fermi level at $x = 0$ microns and $x = 5$ microns.
    Take $n_i = 10^{16}$ m^{-3}, $E_g = 1.1$ eV, $T = 300$ K, and $N_c = N_v$.

5. Find the value of the built in electric field at $x = 0$ if non-uniform n-type doping varies as $N_0 \exp -x^2/L^2$, where $N_0 = 10^{25}$ m^{-3} and $L = 2 \times 10^{-6}$ m, using the parameters for silicon at 300 K. Sketch the band diagram.

6. Draw the band diagram for uniformly doped p-type germanium, with $N_A = 10^{22}$, $N_D = 10^{21}$ m^{-3}, with a drift current density of 3×10^6 A m^{-2}.

7. Sketch the band diagram of a depletion region in the region $0 < x < d$ in a p-type semiconductor, assuming $V = 0$ at $x = 0$ and that there is zero electric field at $x = d$. Does the region curve upwards or downwards?

8. Sketch the band diagram for a pn junction in equilibrium, assuming equal doping on each side. Pay attention to the curvature of the bands in the junction region. Draw the corresponding charge distribution in the device.

9. Repeat (8) for an n^+p junction.

10. Sketch the band diagram for a pn^-n^+ three-layer device in equilibrium, assuming that the central part lies entirely within the depletion region.

5 Non-equilibrium semiconductors

So far we have considered materials and junctions at or close to equilibrium conditions, so that the electron and hole number densities followed the equilibrium equations and the probability of a state being occupied obeyed the Fermi–Dirac distribution law. A flow of energy or information in an electronic system requires that it be disturbed from equilibrium in a controlled way. This chapter is concerned with three issues:

1.   How to change the mobile carrier densities in a semiconductor – a process called *carrier injection*.
2.   How long the material will take to return to equilibrium once the disturbance is removed. This is most important in determining the maximum switching speed of a device.
3.   How far excess mobile carriers will travel within a device: the carrier *diffusion length*. This property determines the maximum physical dimensions of certain types of device.

5.1 Direct and indirect recombination

There are two kinds of equilibrium: static equilibrium, in which an object remains indefinitely in the same position unless disturbed, and dynamic equilibrium, where a constant condition is maintained by balancing inflow and outflow. For example, a brick resting squarely on a flat surface is in static equilibrium, but water molecules evaporating from a liquid surface can be in dynamic equilibrium with molecules condensing from the vapour. Conduction band electrons and valence band holes are in dynamic equilibrium, as electron–hole pairs are generated by electrons jumping up to the conduction band and are lost when they fall down again. Electron–hole pair generation takes place at a constant rate in equilibrium, which depends solely on the temperature in a given material. We shall call this generation rate $\bar{G}$ electron–hole pairs $m^{-3} s^{-1}$ where the bar denotes it is an equilibrium quantity.

The reverse process to generation is called *recombination*, and in equilibrium its value will be $\bar{R}$ electron–hole pairs $m^{-3} s^{-1}$. Then by definition:

$$\bar{G} = \bar{R} \text{ pairs } m^{-3} s^{-1} \tag{5.1}$$

Recombination can occur in two different ways, known as *direct* and *indirect* recombination. Direct recombination is illustrated in Figure 5.1: a free electron falls spontaneously into a vacant hole site, and the band gap energy is surrendered as an electromagnetic pulse of energy – a photon – with a characteristic frequency as given by equation (3.4). Direct recombination is the reverse process of electron–hole pair generation by the absorption of photons.

Indirect recombination occurs through an intermediate state, known as a *trap* or a *recombination centre*. Traps, which may be impurity atoms or defects in the crystal structure, introduce localized energy levels within the band gap. Assuming such a level is initially empty, an electron may be detained there for a short while (Figure 5.2a) and may subsequently fall into a vacant hole (Figure 5.2b). While held in the trap the electron interacts strongly with the thermal energy vibrations (called phonons) of the crystal and yields its energy a little at a time rather than in one packet.

Common semiconductor materials tend to display either one or the other type of recombination behaviour. Thus silicon and germanium are called indirect gap materials, while gallium arsenide and its alloys are direct gap semiconductors.

This division is because the transition between conduction and valence bands requires a matching of both energy and momentum. A direct gap material requires very little momentum shift and a large energy shift, which ties in well

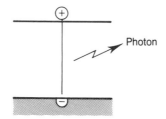

Figure 5.1 Direct recombination of an electron and a hole, with the release of a photon

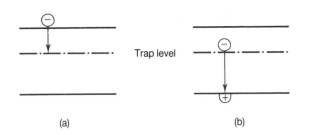

(a)   (b)

Note:
The electron resides in a trap level before recombining. The energy shed goes to heat the crystal lattice.

Figure 5.2 Indirect recombination

with the properties of a photon. Direct recombination is fast and spontaneous and can be likened to a stone falling straight down a cliff. Indirect recombination requires a large shift in both energy and momentum, as if a particle had to jump from a high building and land on a moving bus, James Bond style. Phonons carry little energy but they do carry momentum, so a trap which allows an electron to emit many phonons is the most likely route for indirect recombination.

This distinction has a strong practical significance: any semiconductor device which *emits* light must be made from a direct gap material. Either type of material can absorb light with the creation of electron–hole pairs, but the absorption efficiency of an indirect gap semiconductor is low for photon energies close to the band gap energy.

In the next two sections we shall calculate the number of electrons and holes in direct and indirect gap semiconductors, in which the equilibrium generation rate $\bar{G}$ is supplemented by an extra generation rate $\hat{G}$ pairs m^{-3} s^{-1} uniformly throughout their volume. The hat ($\hat{\ }$) will denote departures from the equilibrium value for recombination rates and charge number densities as well. The total value of a quantity is thus the sum of its equilibrium and excess values, for example:

$$G = \bar{G} + \hat{G} \text{ m}^{-3}\text{s}^{-1}$$
(5.2)

5.1.1 Direct gap semiconductor

The recombination rate is proportional to the number density of electrons and the number density of holes in a direct gap material. This is because the chance of any one electron meeting a hole is proportional to the number of holes, and the loss rate of all electrons will depend on the number of electrons present. Therefore:

$$R = rpn \text{ m}^{-3}\text{ s}^{-1}$$
(5.3)

where r $\mathrm{m^3\,s^{-1}}$ is the recombination coefficient. As an aside we note that this gives a physical meaning to equation (3.24) ($pn = n_i{}^2$). Using equation (5.1):

$$\bar{p}\bar{n} = \bar{G}/r \qquad (5.4)$$

which is a constant at constant temperature. If more holes, for example, are created by doping then the increased chance of electron loss by recombination lowers the electron number density.

Suppose now that uniform irradiation by light increases the electron–hole pair generation rate by $\hat{G}$ pairs $\mathrm{m^{-3}\,s^{-1}}$. The number densities of electrons and holes, and hence the recombination rate, is raised until a *steady state* condition is reached. This means that quantities no longer change with time, *as long as the perturbation remains constant*. The term equilibrium is reserved for unchanging, unperturbed systems.

The total recombination rate becomes:

$$R = \bar{R} + \hat{R} = r(\bar{p} + \hat{p})(\bar{n} + \hat{n}) \ \mathrm{m^{-3}\,s^{-1}} \qquad (5.5)$$

This can be simplified by substituting:

$$\bar{R} = r\bar{p}\bar{n} \ \mathrm{m^{-3}\,s^{-1}} \qquad (5.6)$$

the equilibrium condition and:

$$\hat{n} = \hat{p} \ \mathrm{m^{-3}} \qquad (5.7)$$

because electrons and holes are formed in pairs. We note also that the total generation and recombination rates are equal in the steady state:

$$\bar{G} + \hat{G} = \bar{R} + \hat{R} \ \mathrm{m^{-3}\,s^{-1}} \qquad (5.8)$$

hence using equation (5.1):

$$\hat{G} = \hat{R} \ \mathrm{m^{-3}\,s^{-1}} \qquad (5.9)$$

Thus equation (5.5) becomes:

$$\hat{n}^2 + \hat{n}(\bar{p} + \bar{n}) = \frac{\hat{G}}{r} \ \mathrm{m^{-6}} \qquad (5.10)$$

This is the full solution and shows that the excess electron or hole density depends on the excess generation rate, the material type and its doping in a non-linear fashion. Some simpler approximations are often valid:

1. *Low level injection* ($\hat{n} \ll \bar{n}$ or $\hat{n} \ll \bar{p}$): in this case the $\hat{n}^2$ term in equation (5.10) can be neglected to give:

$$\hat{n} = \frac{\hat{G}}{r(\bar{p} + \bar{n})} \ \mathrm{m^{-3}} \qquad (5.11)$$

2. *Low level injection*, p-*type material* ($\bar{p} \gg \bar{n}$): equation (5.11) becomes:

$$\hat{n} = \frac{\hat{G}}{r\bar{p}} \tag{5.12}$$

and we can often approximate:

$$p = \bar{p} + \hat{p} \simeq \bar{p}$$
$$n = \bar{n} + \hat{n} \simeq \hat{n} \tag{5.13}$$

3. *Low level injection,* n-*type material* ($\bar{n} \gg \bar{p}$): equation (5.11) approximates to:

$$\hat{n} = \frac{\hat{G}}{r\bar{n}} \tag{5.14}$$

and the carrier number densities are:

$$p = \bar{p} + \hat{p} \simeq \hat{p}$$
$$n = \bar{n} + \hat{n} \simeq \bar{n} \tag{5.15}$$

In cases (2) and (3) note that the quantities $r\bar{p}$ and $r\bar{n}$ have the units s^{-1}. The reciprocals of these are the characteristic *lifetimes* of the minority species in each case. Low level injection strongly influences the number density of minority carriers, but by definition alters the majority carrier density very little.

EXAMPLE

■ In one semiconductor material doped p-type with $N_A = 10^{20}$ m^{-3} the recombination coefficient is 5×10^{-12} m^3 s^{-1} and $n_i = 10^{16}$ m^{-3}. Find the minority carrier lifetime, the equilibrium generation rate, and find the electron and hole number densities if light causes an excess generation of 10^{22} pairs m^{-3} s^{-1}.

□ The material is p-type with $N_A \gg n_i$, so:

$$\bar{p} \simeq N_A = 10^{20} \text{ m}^{-3}$$
$$\bar{n} \simeq n_i^2/N_A = 10^{12} \text{ m}^{-3}$$

The minority carrier lifetime (τ_e) is

$$\tau_e = 1/r\hat{p}$$
$$= 2 \times 10^{-9} \text{ s}$$

The equilibrium generation rate $\bar{G}$ is found by:

$$\bar{G} = \bar{R} = r\bar{p}\bar{n}$$

therefore $\bar{G} = 5 \times 10^{20}$ m^{-3} s^{-1}.

Assuming the light causes only low level injection:

$$\hat{n} = \hat{p} = \hat{G}/r\bar{p}$$

$$= 2 \times 10^{13} \text{ m}^{-3}$$

The assumption is verified, because $\hat{n} \ll \bar{p}$.

Hence the number densities can be found:

$$n = \bar{n} + \hat{n} = 10^{12} + 2.10^{13}$$

$$= 2.1 \times 10^{13} \text{ m}^{-3} \ (\simeq \hat{n})$$

$$p = \bar{p} + \hat{p} = 10^{20} + 2.10^{13}$$

$$\simeq 10^{20} \text{ m}^{-3} \ (= \bar{p})$$

So low level injection has increased the minority electron number density by a factor of 20, while hardly disturbing the majority hole number density.

(Note that the injection was still low level even though $\hat{G} \gg \bar{G}$ in this case. This is examined further in Exercise 6(b).)

5.1.2 Indirect gap semiconductor

Because recombination occurs here through an intermediate trap state, the recombination rate depends on the following:

1. N_t, the number density of trap sites.
2. c_n and c_p (m³s⁻¹), the capture coefficients for electrons and and holes respectively, which are a measure of the trap's ability to catch hold of carriers, and are akin to the direct recombination coefficient r.
3. E_t, the trap energy level in the band gap, which will determine the likelihood of the level being filled or empty in equilibrium.

Three cases of trap energy level are worth considering, as shown in Figure 5.3:

1. The trap states with energy E_{t1} are close to the conduction band and above the Fermi level, so will tend to be empty. Mobile electrons will have little difficulty entering the trap states but must shed a good deal of energy to transfer to the valence band. It is more likely that they will be re-emitted into the conduction band.

2. Traps at energy E_{t2} are near the middle of the band gap. They will tend to be empty in p-type material $(E_{t2} > E_F)$ and full in n-type material

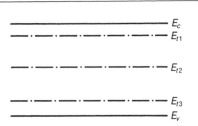

Note:
Traps with energy E_{t1} can capture electrons easily, but not holes; at E_{t2} either can be captured; at E_{t3} holes are captured easily, but not electrons. Levels near the gap centre make better recombination centres.

Figure 5.3 Trap energy levels

$(E_{t2} < E_F)$, so will tend to be in the correct condition to receive minority carriers. Having captured a minority carrier (an electron in p-type material, for example), the energy change to reach either band is similar, so the probability of re-emission into the minority carrier band is comparable with that of recombination with a majority carrier.

3.   Traps with energies E_{t3} near the valence band will always tend to be full. Because of their position it is far more likely that they might empty themselves by dropping an electron into a valence band hole (a process also called hole capture) than by emitting an electron into the conduction band.

From the above discussion it is clear that indirect recombination is most efficient via states near the middle of the band gap. Recombination can be viewed as the alternate capture of a conduction band electron and a valence band hole by a trap. Traps near the band edges perform one of these functions well, but do the other extremely poorly: traps near the gap centre are the best compromise between the two. For this reason sites near the band edges which temporarily hold and re-emit carriers into the same sites are sometimes called traps, while band gap centre sites are referred to as recombination centres.

Although the mathematics of indirect recombination is a little more complex, simple results emerge subject to the following assumptions:

1. $n \gg n_i$ or $p \gg n_i$ (i.e. doped material).
2. $E_t = E_{Fi}$ (i.e. trap near the middle of the gap).
3. $c_n = c_p = c$ (equal electron and hole capture coefficients).
4. $\hat{n} \ll \bar{n}$ or $\hat{n} \ll \bar{p}$ (low level injection).

Under these circumstances recombination is dominated by the number of minority carriers and the number of recombination centres. It does not depend on the number of majority carriers, unlike direct recombination, because it is assumed that one can be trapped readily, once a minority carrier has been trapped.

1. n-type material ($\bar{n} \gg \bar{p}$): holes are the minority carriers. The equilibrium condition is:

$$\bar{G} = \bar{R} = cN_t\bar{p} \ \mathrm{m}^{-3}\,\mathrm{s}^{-1} \tag{5.16}$$

If excess generation of electron–hole pairs takes place at $\hat{G}$ pairs m^{-3} s^{-1}, the excess recombination rate must balance this in the steady state:

$$\hat{G} = \hat{R} = cN_t\hat{p} \ \mathrm{m}^{-3}\,\mathrm{s}^{-1} \tag{5.17}$$

The product cN_t has the units of s^{-1}, so its inverse is identified with the hole minority lifetime τ_h:

$$\tau_h = 1/cN_t \ \mathrm{s} \tag{5.18}$$

2. p-type material ($\bar{p} \gg \bar{n}$): here electrons are the minority carriers, so the equilibrium recombination rate depends on the chance of an electron being captured by a recombination centre. Once trapped the large number of holes make recombination almost certain, so the hole density is unimportant. In equilibrium:

$$\bar{G} = \bar{R} = cN_t\bar{n} \ \mathrm{m}^{-3}\,\mathrm{s}^{-1} \tag{5.19}$$

And if an excess generation $\hat{G}$ m^{-3} s^{-1} is added, the steady state condition is:

$$\hat{G} = \hat{R} = cN_t\hat{n} \tag{5.20}$$

and the electron minority lifetime τ_e is:

$$\tau_e = 1/cN_t \ \mathrm{s} \tag{5.21}$$

EXAMPLE

■   A piece of indirect gap semiconductor, with $n_i = 10^{16}$ m^{-3} and doped uniformly n-type with $N_D = 10^{21}$ m^{-3} has a minority carrier lifetime of 2 microseconds. It is bombarded with high energy particles which create extra electron–hole pairs at a rate of 10^{18} m^{-3} s^{-1}. Find the total generation rate and the number densities of electrons and holes (1) in equilibrium; (2) in the steady state.

☐   1.   Equilibrium implies that the material is free from bombardment. Equation (5.18) or (5.21) supply a value for the product cN_t:

$$cN_t = 1/\tau_h$$
$$= 5 \times 10^5 \text{ s}^{-1}$$

τ_h has been chosen as the minority carrier lifetime because the material is n-type. Since $N_D \gg n_i$, the equilibrium values $\bar{n}$ and $\bar{p}$ are:

$$\bar{n} \simeq N_D = 10^{21} \text{ m}^{-3}$$
$$\bar{p} \simeq n_i^2/N_D = 10^{11} \text{ m}^{-3}$$

These values of cN_t and $\bar{p}$ can be used in equation (5.16) to find the equilibrium generation (or recombination) rate:

$$\bar{G} = cN_t\bar{p}$$
$$= 5 \times 10^{16} \text{ m}^{-3}$$

2. The steady state condition means that the bombardment has been going on sufficiently long for a new balance to be struck, so that n and p are virtually constant again. The total generation rate is the sum of the equilibrium rate and the additional rate caused by bombardment:

$$G = \bar{G} + \hat{G}$$
$$= 5.10^{16} + 10^{18}$$
$$= 1.05 \times 10^{18} \text{ m}^{-3}\text{s}^{-1}$$

The additional minority carrier density is found from equation (5.17):

$$\hat{p} = \hat{G}/cN_t$$
$$= 10^{18}/5 \times 10^5$$
$$= 2.10^{12} \text{ m}^{-3}$$

This equals $\hat{n}$, the additional majority carrier density. The total free carrier densities are:

$$n = \bar{n} + \hat{n}$$
$$= 10^{21} + 2.10^{12}$$
$$\simeq 10^{21} \text{ m}^{-3} \quad (\simeq \bar{n})$$
$$p = \bar{p} + \hat{p}$$
$$= 10^{11} + 2.10^{12}$$
$$= 2.1 \times 10^{12} \quad (\simeq \hat{p})$$

As in the example using direct recombination, low level injection has a marked effect on the minority carrier density, while leaving the majority carrier density virtually unchanged. Note also that the low level injection condition ($\hat{n} \ll \bar{n}$ or $\hat{n} \ll \bar{p}$) has been satisfied, although the perturbation $\hat{G}$ was greater than the equilibrium value $\bar{G}$.

5.2 Minority carrier decay in time

In the last section it has been argued that constant low level injection of electron–hole pairs in a semiconductor will add a small number of extra carriers to each band. In each case the minority carrier density was more strongly perturbed, and the loss rate of minority carriers determined the amount of disturbance. Here we shall consider what happens to a semiconductor which has been constantly perturbed when the outside influence suddenly ceases, at a time $t = 0$. Clearly the material will tend to return to equilibrium; but how long will this take?

Suppose that the *excess* carrier densities at $t = 0$ are $\hat{n}_0$ ($= \hat{p}_0$), which can be found from steady state calculations. During the decay phase $\hat{G} = 0$, and the generation rate and recombination rate are out of balance:

$$\bar{G} \neq \bar{R} + \hat{R} \tag{5.22}$$

For both direct and indirect gap semiconductors the excess recombination rate $\hat{R}$ depends on the number of excess minority carriers and their characteristic lifetimes.

For a p-type semiconductor:

$$\hat{R} = \hat{p}/\tau_h \ \mathrm{m^{-3}\,s^{-1}} \tag{5.23}$$

where $\hat{p}$ varies with time now. In any short time interval of length dt, the number of minority carriers lost per unit volume is $(\hat{p}/\tau_h)dt$, which may be expressed as:

$$d\hat{p} = -(\hat{p}/\tau_h)dt \ \mathrm{m^{-3}} \tag{5.24}$$

The minus sign indicates that $\hat{p}$ has gone down by an amount $d\hat{p}$ in the time interval dt. Equation (5.24) can be written:

$$\frac{d\hat{p}}{\hat{p}} = \frac{-dt}{\tau_h} \tag{5.25}$$

which can be integrated. The limits are the initial conditions ($\hat{p} = \hat{p}_0$ at $t = 0$) and the value of $\hat{p}$ at time t, which we wish to find:

$$\int_{\hat{p}_0}^{\hat{p}} \frac{d\hat{p}}{\hat{p}} = \int_0^t \frac{-dt}{\tau_h} \tag{5.26}$$

$$\ln \frac{\hat{p}}{\hat{p}_0} = \frac{-t}{\tau_h} \tag{5.27}$$

or:

$$\hat{p} = \hat{p}_0 \exp\frac{-t}{\tau_h} \qquad (5.28)$$

This expression is plotted in Figure 5.4. Note that it is the excess carriers which die away, leaving the equilibrium number density. It seems that the material never returns to equilibrium, because the exponential only has a zero value at infinite time. This is not so, because once its value drops below that of (1/semiconductor volume) there is less than one excess carrier remaining. Electronic engineers are usually satisfied to wait until $\hat{p}$ has decayed to 10% or maybe 1% of its initial value, which takes $2.3\tau_h$ or $4.6\tau_h$ respectively.

The minority carrier lifetime is a good example of the scale of a process, by which we mean the characteristic length or time associated with the major part of the change. When considering the switching of semiconductor devices, τ_e or τ_h are often the timescales of interest because they control the rate at which perturbed minority carrier densities return to equilibrium through recombination. If there are other timescales – for example there may be one for the removal of excess charges by the flow of electric current – then the shortest timescale generally dominates.

The minority carrier lifetime in a direct gap semiconductor such as GaAs is usually 1 ns or less, while a minority carrier lifetime in an indirect gap semiconductor may be up to 100 μs (in silicon) or even several milliseconds (in germanium). The minority lifetime in silicon can be reduced to around 1 ns by doping it with gold atoms, which introduce trapping levels in the band gap (Figure 5.5). The more gold atoms there are, the higher is the value for N_t and

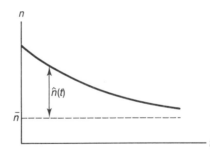

Figure 5.4 The exponential decay in time of excess minority carriers

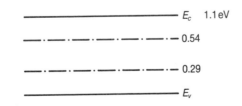

Figure 5.5 Levels in the gap introduced by gold doping

the lower the lifetime. There is a penalty to pay, however: the staircase of trap levels will cause extra generation of minority carriers and hence a larger leakage current in some devices designed to block current.

EXAMPLE

■ In n-type germanium with $n_i = 2.5 \times 10^{19}$ m^{-3} and $N_D = 10^{22}$ m^{-3}, a flash of light doubles the hole number density. Calculate the time taken for the hole density to decay to 8×10^{16} m^{-3} if $\tau_h = 3$ ms.

□ First calculate $\bar{p}$:

$$\bar{p} \simeq n_i^2/N_D = 6.25 \times 10^{16} \text{ m}^{-3}$$

The light flash doubles this, so the initial excess hole density equals $\bar{p}$:

$$\hat{p}_0 = 6.25 \times 10^{16} \text{ m}^{-3}$$

At some time t later, the excess has decayed to:

$$\hat{p} = 8 \times 10^{16} - 6.25 \times 10^{16}$$

$$= 1.75 \times 10^{16} \text{ m}^{-3}$$

Equation (5.27) can be recast to find t for this value of p:

$$t = \tau_h \ln \frac{\hat{p}_0}{\hat{p}}$$

$$= 3 \times 10^{-3} \ln (6.25/1.75)$$

$$= 3.82 \text{ ms}$$

5.3 Minority carrier decay in space

Figure 5.6 shows the next case to be considered. Low level injection – by light or some other means within a device – keeps the excess minority carrier density constant at some position $x = 0$ within the material. For positive values of x there is only equilibrium generation of electron–hole pairs. Disregarding any possible charging effects the excess minority carriers will tend to diffuse to the right, into the unperturbed region where the minority number density is lower. As they diffuse they will recombine until at some distance away the material remains close to equilibrium.

Figure 5.7 shows how this spatial decay can be calculated. Just as a small time interval dt at a general time t was considered in the last section, so here a

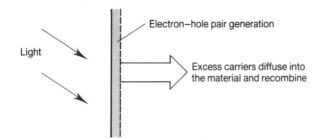

Figure 5.6 Excess carriers can be created in a narrow region close to the surface

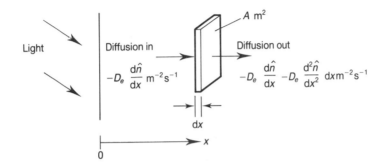

Note:
The difference between the numbers diffusing in and out is the number lost by recombination within the slab.

Figure 5.7 Calculating spatial decay

small volume of thickness dx at a distance x from the replenished surface will be used. It is assumed to have an area A m^2 into the page. A profit and loss balance of excess minority carriers can be reckoned for this volume. The equilibrium generation and recombination of carriers takes place as well, but these cancel out. Three other elements must be considered:

1. Diffusion of excess minority carriers into the volume. Suppose that they are electrons, with a number density $\hat{n}$ which will vary with position x. From Fick's law (equation 2.18):

$$\text{number of electrons diffusing in at } x = -AD_e\frac{d\hat{n}}{dx} \text{ electrons s}^{-1} \quad (5.29)$$

2. The number of excess minority carriers recombining in the volume:

$$\text{number lost through recombination} = \hat{n}A\,dx/\tau_e \text{ electrons s}^{-1} \quad (5.30)$$

where the number density (number per unit volume) has been multiplied by the slab volume to give the number of excess electrons, and division by the lifetime gives the number lost per second.

3.   Diffusion of excess minority carriers out of the volume. To calculate this it
     is necessary to find the value of the diffusion flux at the position $(x + dx)$.
     A function $y(x)$ which has the value y at position x takes the value:

$$y(x + dx) = y + \frac{dy}{dx} dx \tag{5.31}$$

at position $x + dx$. In this case the function is given by equation (5.29), so
equation (5.31) gives:

number of electrons diffusing out at $(x + dx)$

$$= -AD_e\left(\frac{d\hat{n}}{dx} + \frac{d^2\hat{n}}{dx^2}dx\right) \text{ electrons s}^{-1} \tag{5.32}$$

A balance can now be struck using equation (5.29), (5.30) and (5.32):

$$-AD_e\frac{d\hat{n}}{dx} = \frac{\hat{n}Adx}{\tau_e} - AD_e\frac{d\hat{n}}{dx} - AD_e\frac{d^2\hat{n}}{dx^2}dx \tag{5.33}$$

which simplifies to:

$$\frac{d^2\hat{n}}{dx^2} - \frac{\hat{n}}{D_e\tau_e} = 0 \tag{5.34}$$

This is a second order linear differential equation, which has a standard
solution with A_1 and A_2 as arbitrary constants:

$$\hat{n} = A_1\exp\frac{-x}{\sqrt{(D_e\tau_e)}} + A_2\exp\frac{+x}{\sqrt{(D_e\tau_e)}} \tag{5.35}$$

The constants can be found by applying boundary conditions:

1. $\hat{n} = \hat{n}_0$ at $x = 0$ (the replenished boundary)

2. $\hat{n} = 0$ as $x \to \infty$ (equilibrium at a distance)

These give:

$$A_1 = \hat{n}_0 \tag{5.36}$$

$$A_2 = 0$$

or

$$\hat{n} = \hat{n}_0\exp\frac{-x}{\sqrt{(D_e\tau_e)}} \text{ m}^{-3} \tag{5.37}$$

This result shows that the number density of excess minority carriers dies away
exponentially from the replenished surface. It will be used again when calculating the diode I/V characteristic. The scale length of the decay is $\sqrt{(D_e\tau_e)}$ (check
the units) and is called the *diffusion length* L_e for minority electrons:

$$L_e = \sqrt{(D_e \tau_e)} \text{ m} \tag{5.38}$$

The hole diffusion length L_h is defined similarly as $\sqrt{(D_h \tau_h)}$. Its magnitude for electrons is typically 3 mm in germanium, 0.2 mm in silicon, and 2 μm in gallium arsenide. These figures have had a profound influence on the development of the semiconductor industry, because a major class of devices, bipolar devices, depend on making the heart of the devices much smaller than L_e or L_h. This is easiest to achieve in germanium, which dominated the early transistor market. Silicon has other advantages, notably a stable insulating oxide, and currently holds the largest market share. Bipolar gallium arsenide transistors are still in the laboratory (1990).

EXAMPLE

■ The hole number density is raised to 10^{14} m^{-3} at the surface of an n-type silicon sample at 300 K ($n_i = 10^{16}$ m^{-3}, $N_D = 10^{20}$ m^{-3}, $\mu_h = 0.05$ m^2 V^{-1} s^{-1}, $\tau_h = 3$ μs). Assuming that no excess carrier generation occurs away from the surface, find the depth at which the hole density falls to 10^{13} m^{-3}.

☐ First find $\bar{n}$ and $\bar{p}$. Since $N_D \gg n_i$:

$$\bar{n} \simeq N_D = 10^{20} \text{ m}^{-3}$$

$$\bar{p} \simeq n_i^2/N_D = 10^{12} \text{ m}^{-3}$$

The excess hole density at the surface, $\hat{p}_0$, is therefore:

$$\hat{p} = 10^{14} - 10^{12}$$

$$= 9.9 \times 10^{13} \text{ m}^{-3}$$

Where the hole density has decayed to 10^{13} m^{-3}, the excess is:

$$\hat{p} = 10^{13} - 10^{12}$$

$$= 9 \times 10^{12} \text{ m}^{-3}$$

The hole diffusion length is required, so the hole diffusion coefficient D_h must be found from the Einstein relation:

$$D_h = \mu_h \frac{kT}{q}$$

$$= 1.29 \times 10^{-3} \text{ m}^2 \text{ s}^{-1}$$

Hence L_h is found:

$$L_h = \sqrt{(D_h \tau_h)}$$

$$= 6.2 \times 10^{-5} \text{ m}$$

Rearranging equation (5.37) for the case of holes, the depth x at which this occurs is:

$$x = L_h \ln \frac{\hat{p}_0}{\hat{p}}$$

$$= 1.49 \times 10^{-4} \text{ m}$$

5.4 Surface recombination

So far it has been assumed that surfaces are merely convenient planes to shine light into a semiconductor and thereby inject minority carriers. The physicist who said 'God made solids, but the devil made surfaces' finds a ready echo among electronic device engineers, because the point at which a crystal structure ends almost inevitably introduces a large number of levels within the band gap. These surface states are powerful recombination centres, the density of which can be defined as N_{ts} traps *per unit area*. Multiplying by a mean capture coefficient c m^{-3} s^{-1} for the traps gives a quantity S:

$$S = cN_{ts} \text{ m s}^{-1} \tag{5.39}$$

which is called the *surface recombination velocity*. A good deal of device engineering is concerned with 'passivating' semiconductor surfaces and reducing S. The success of silicon devices arises from the excellent passivation possible by growing a silicon dioxide layer on the silicon surface. Failure to do this will lead to the device performing well below the characteristics expected from its ideal bulk properties. The recombination rate per unit area at the surface $\hat{R}_s$ is then:

$$\hat{R}_s = S\hat{p} \text{ pairs m}^{-2} \text{ s}^{-1} \tag{5.40}$$

where $\hat{p}$ is evaluated at the surface.

Summary

Main symbols introduced

Symbol	Unit	Description
c, c_n, c_p	m^3 s^{-1}	capture coefficient for electrons, holes
E_t	J	trap energy level
G	m^{-3} s^{-1}	electron–hole pair generation rate

Symbol	Unit	Description
L_e, L_h	m	electron, hole diffusion lengths
N_t	m^{-3}	trap number density
N_{ts}	m^{-2}	surface state density
r	$m^3\,s^{-1}$	electron–hole direct recombination coefficient
R	$m^{-3}\,s^{-1}$	electron–hole pair recombination rate
S	$m\,s^{-1}$	surface recombination velocity
τ_e, τ_h	s	electron, hole minority carrier lifetime.
$^{-}$	superscript	equilibrium value
$^{\wedge}$	superscript	excess above equilibrium value

Main equations

$$R = rpn \text{ m}^{-3}\text{s}^{-1} \quad \text{(direct recombination rate)} \tag{5.3}$$

$$R = cN_t x \text{ m}^{-3}\text{s}^{-1} \quad (x: \text{minority density} - \text{indirect recombination}) \tag{5.16}$$

	Minority carrier lifetime	
	τ_e (p-type)	τ_h (n-type)
Direct gap	$1/r\bar{p}$	$1/r\bar{n}$
Indirect gap	$1/c_n N_t$	$1/c_p N_t$

$$\hat{p} = \hat{p}_0 \exp -t/\tau_h \text{ m}^{-3} \quad \text{(minority carrier decay in time)} \tag{5.28}$$

$$\hat{n} = \hat{n}_0 \exp -x/L_e \text{ m}^{-3} \quad \text{(minority carrier decay in space)} \tag{5.37}$$

$$L_e = \sqrt{(D_e \tau_e)} \text{ m} \quad \text{(hole diffusion length)} \tag{5.38}$$

Exercises

1. The lifetime of electrons or holes in intrinsic GaAs is 10 ns, and $n_i = 10^{13}$ m^{-3}. Find the equilibrium thermal generation rate of electron–hole pairs and the recombination coefficient. Find the generation rate $\bar{G}$ and the electron lifetime if it is doped with 10^{18} m^{-3} acceptors.

2. Gallium arsenide ($E_g = 1.43$ eV) is illuminated with light of wavelength 0.83 μm at an intensity of 2 W m^{-2}. If it is absorbed uniformly in a 0.3 μm thick layer, find n and p within the layer. Take $r = 10^{-5}$ m^3 s^{-1}, $n_i = 10^{13}$ m^{-3}, $N_D = 10^{18}$ m^{-3}.

3. In a direct gap semiconductor $r = 5.10^{-12}$ m^3 s^{-1}, $n_i = 10^{16}$ m^{-3} and $N_D = 10^{19}$ m^{-3}. Find $\bar{G}$ and the minority carrier lifetime.

4. A direct gap semiconductor is doped with 10^{17} m^{-3} donors and 2×10^{16} m^{-3} acceptors, with $n_i = 10^{16}$ m^{-3} and $r = 10^{-11}$ m^3 s^{-1}. Light creates extra carrier pairs at a rate 10^{23} m^{-3} s^{-1}. Find the electron and hole number densities. Is this low level injection?

5. Steady illumination raises the carrier densities in an indirect gap semiconductor to $n = 2.09 \times 10^{18}$ m^{-3}, $p = 1.00 \times 10^{22}$. Find the equilibrium and excess carrier generation rate if $\tau_e = 2$ ms, $n_i = 3 \times 10^{19}$ m^{-3}, and $N_A = 10^{22}$ m^{-3}.

6. (a) Show that low level injection in a direct or an indirect gap n-type semiconductor produces $(\hat{G}\bar{p}/\bar{G})$ excess minority carriers.
   (b) Using the binomial expansion of the full quadratic solution, show that the low level injection condition in a p-type direct gap semiconductor requires that $(\hat{G}/\bar{G}) \ll (\bar{p}/\bar{n})$.

7. Conductivity modulation is the name given to the change in conductivity caused by excess minority carriers. Find the percentage change in conductivity in p^- doped silicon ($N_A = 5.10^{16}$ m^{-3}, $n_i = 10^{16}$ m^{-3}) if the minority carrier lifetime is 2 μs and 10^{21} m^{-3} s^{-1} extra electron–hole pairs are injected. Take $\mu_e = 0.1$ and $\mu_h = 0.05$ m^2 V^{-1} s^{-1}.

8. Minority carrier injection suddenly ceases at $t = 0$ in the semiconductor of (7). Find the time taken for the conductivity to return to within 0.1% of its equilibrium value.

9. The width of the base region of a bipolar transistor should be at least ten times smaller than the minority carrier diffusion length for efficient operation. Find the maximum permissible level of gold doping in a p-type silicon base region 4 μm wide if the electron capture coefficient is 10^{-14} m^3 s^{-1}. Assume $T = 300$ K, $\mu_e = 0.14$ m^2 V^{-1} s^{-1}.

10. Excess carrier generation occurs uniformly throughout a p-type semiconductor at a rate $\hat{G}$ m^{-3} s^{-1}, for $x > 0$, and there is a surface with a recombination velocity S at $x = 0$. Show that $\hat{n}$ satisfies the equation:

$$\frac{d^2\hat{n}}{dx^2} - \frac{\hat{n}}{D_e \tau_e} = \frac{-\hat{G}}{D_e}$$

Solve this equation subject to the boundary conditions:

(a) $d\hat{n}/dx \rightarrow 0$ as $x \rightarrow \infty$

(b) diffusion flux at $(x = 0)$, $= -D_e d\hat{n}/dx = -S\hat{n}(0)$ to obtain the spatial variation of n:

$$\hat{n} = \hat{G}\tau_e - \left(\frac{S\tau_e L_e \hat{G}}{D_e + L_e S}\right)\exp(-x/L_e)$$

6 | The *pn* junction diode – dc

6.1 Band diagram

A *pn* junction diode is formed in a semiconductor at the boundary between a *p*-type region and an *n*-type region. It is not sufficient just to join together two separate pieces of semiconductor because the interface will possess a high recombination velocity, which will swamp any useful effects. The *p*- and *n*-type regions must be formed by varying the doping within a single semiconductor crystal. For example, in Figure 6.1a N_D is greater than N_A on the left, and less on the right. The cross-over point is called the metallurgical junction. The effective doping will therefore be *n*-type on the left and *p*-type on the right, creating the junction. This type of doping creates a linear junction.

In the following discussion it will be assumed that the doping levels are constant on either side of the metallurgical junction, where the material switches suddenly from being strongly *n*-type to being firmly *p*-type. This is called an abrupt junction (Figure 6.1b). If the doping is far heavier on one side the junction is single-sided abrupt.

We will now consider the band diagram of the *pn* junction, to arrive at a

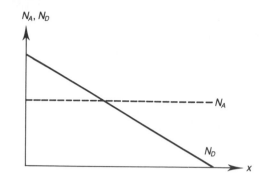

(a) Doping in a linear junction

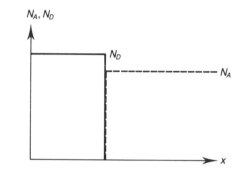

(b) An abrupt junction

Figure 6.1 Doping in junctions

qualitative idea of its I/V characteristic. The equilibrium band diagram was derived in Chapter 4. Unequal doping causes a strong diffusion current to flow, which moves charges out of their home regions. The *n*-type side loses electrons, gains holes and acquires a net positive charge, while the *p*-type side becomes charged negative. These extra charges are not uniformly distributed over the whole material but are concentrated near the metallurgical junction in a depletion region, as shown in Figure 6.2a, because of the powerful electrostatic attraction between the opposite charges. The band diagram associated with this charge distribution is drawn in Figure 6.2b. The Fermi level is constant because the junction is in equilibrium. The charged regions correspond to the curved junction region, upwards curvature for the negatively charged region and downwards for the positive.

The flat energy levels on either side of the depletion region mean that no current is flowing, the doping is uniform and the regions are electrically neutral. The raising of the electron energy levels on the *p* side corresponds to that side

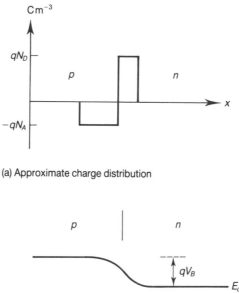

(a) Approximate charge distribution

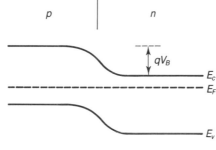

(b) Equilibrium band diagram

Figure 6.2 Charge distribution and equilibrium band diagram of a *pn* junction

possessing a net negative charge, and hence a more negative potential, than the
n-type side. This potential difference between the two sides is called the built in
potential, V_B. The built in potential forms a barrier to the continued diffusion of
charges away from high concentration regions. A dynamic equilibrium exists at
the junction: a few of the more energetic majority carriers on each side climb up
over the barrier, while an equal number of minority carriers slide back down.

Now suppose that an external potential V_A is applied across the junction
with the *p*-type side made negative (Figure 6.3a). This takes the junction out of
equilibrium, so the Fermi level E_F is no longer constant. E_F is qV_A joules higher
on the *p*-type side, and the other levels rise with it. Does a current flow? The
band diagram in Figure 6.3b reveals that this bias increases the junction barrier
potential, from V_B to $V_A + V_B$ (both V_A and V_B are positive here). The number
of majority carriers on each side with sufficient energy to overcome the barrier
is greatly reduced. Minority carriers close to the junction are looking at a very
favourable slide down to a lower energy and so they continue to flow. This
means that an overall current does flow round the circuit but it is a very small

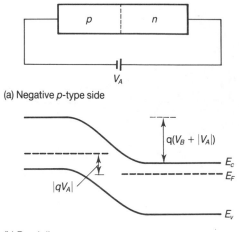

(a) Negative *p*-type side

(b) Band diagram

Figure 6.3 Reverse bias

leakage current, limited by the number of minority carriers diffusing into, or generated near, the junction. An applied potential difference in this sense is called a reverse bias.

Forward bias is with the *p*-type side made positive (Figure 6.4a). Now the applied potential acts to reduce the barrier to $(V_B - V_A)$, where $V_B, V_A > 0$ (Figure 6.4b). The result is like knocking several metres off the height of a reservoir dam: a large current is unleashed as the pent up majority electrons and holes flood down the concentration gradients. This action is quite unlike the

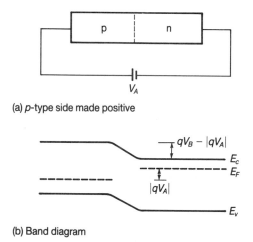

(a) *p*-type side made positive

(b) Band diagram

Figure 6.4 Forward bias

conventional driving of current through a resistance by an electromotive force. The applied bias potential serves mainly to *reduce a hindrance* to the flow of current. A little of the applied potential is used in driving the majority carriers towards the junction, but most of it is used in reducing the barrier.

The band diagram shows that the *pn* junction passes current in one direction only, so it is a rectifier – the electrical equivalent of a car tyre valve. The term diode is inherited from the vacuum diode rectifier, where it is a contraction of di-electrode. These electrodes are two metal plates, one of which – the cathode – is heated red hot to allow electrons to escape into the vacuum, while the anode plate receives them if it is biased positive. Reversing the applied voltage does not cause a current flow, because the cold anode emits no electrons. The terms anode and cathode are also applied to the *pn* junction diode, where the *p*-type side is the anode. Figure 6.5 gives the circuit symbol for the junction diode, and a memory aid.

6.2 The built in potential

The following conventions and notation will be used in calculating V_B and deriving the I/V characteristic:

1. $V_A > 0$ is forward bias, with the *p*-type side made positive.
2. $V_A < 0$ is reverse bias, with the *p*-side negative.

The total barrier potential, V_t is then:

$$V_t = V_B - V_A \text{ V} \tag{6.1}$$

in both forward and reverse bias, if V_B is taken to be positive.

n_n = electron number density on the *n*-type side
n_p = electron number density on the *p*-type side
p_p = hole number density on the *p*-type side
p_n = hole number density on the *n*-type side

Thus n_n and p_p are majority carrier densities, while n_p and p_n are minority carrier densities.

$V_{n,p}$ = intrinsic Fermi potential on the *n*-side, *p*-side

ϕ = Fermi potential

Figure 6.5 The circuit symbol for the *pn* junction, and a memory aid

These quantities are the local potentials, as defined by equations (4.13) and (4.14) $E_{Fi} = -qV$; $E_F = -q\phi$. The total barrier potential V_t is therefore:

$$V_t = V_n - V_p \text{ V} \tag{6.2}$$

as shown in Figure 6.6. The equilibrium equations (4.15) and (4.16) for the majority carrier densities may be written:

$$\bar{n}_n = n_i \exp\frac{q(V_n - \phi)}{kT} \quad (n\text{-side}) \tag{6.3}$$

$$\bar{p}_p = n_i \exp\frac{q(\phi - V_p)}{kT} \quad (p\text{-side}) \tag{6.4}$$

Multiplying these together:

$$\bar{n}_n\bar{p}_p = n_i{}^2 \exp\frac{q(V_n - V_p)}{kT} \tag{6.5}$$

The Fermi level is constant across the device in equilibrium, so is equal in equations (6.3) and (6.4), and the total barrier height V_t is the built in potential V_B. Hence:

$$V_B = \frac{kT}{q}\ln\frac{\bar{n}_n\bar{p}_p}{n_i{}^2} \tag{6.6}$$

Away from the junction region the majority carrier densities take their bulk equilibrium values, so if the doping is not too light:

$$V_B = \frac{kT}{q}\ln\frac{N_A N_D}{n_i{}^2} \tag{6.7}$$

For example, in silicon at 300 K, with $N_A = N_D = 10^{22} \text{ m}^{-3}$, V_B is 0.69 volts.

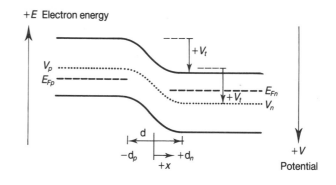

Note.
$V_n - V_p$ equals the total barrier potential in or out of equilibrium.

Figure 6.6 The total barrier potential, V_t

6.3 Minority carrier injection

The equilibrium equations should not strictly be used when a potential V_A is applied, but they can be used with caution provided the injection of minority carriers across the junction remains low level. The term minority carrier injection is applied to a forward biased *pn* junction, because carriers leave the side on which they are the majority species and are injected across the depletion region into a region where they are in the minority.

Further understanding of the barrier is obtained by substituting $n_i^2 = \bar{n}_n \bar{p}_n$ in equation (6.6):

$$V_B = \frac{kT}{q} \ln \frac{\bar{p}_p}{\bar{p}_n} \tag{6.8}$$

or

$$\bar{p}_n = \bar{p}_p \exp \frac{-qV_B}{kT} \tag{6.9}$$

This is none other than the Boltzmann relation for the hole number densities either side of a potential barrier of height V_B volts. A similar expression exists for electrons. The current flow in forward bias is strongly dependent on the number of minority carriers injected. To calculate this for the case of hole injection equation (4.5) is used:

$$J_h = q p \mu_h \mathscr{E} - q D_h \frac{dp}{dx} \, \mathrm{A\,m^{-2}} \tag{6.10}$$

This expresses the hole current density in terms of its drift and diffusion components. J_h is zero across a *pn* junction in equilibrium because of the precise balance between two large current components. An estimate of the hole diffusion current in a diode of area 1 mm^2 and a *p*-side doping of 10^{24} m^{-3} with a depletion region 2 μm wide, and $D_h = 1.2 \times 10^{-3}$ m^2 s^{-1}, is:

$$I_{hdiff} \simeq A q D_h \frac{\Delta p}{\Delta x} = 96 \, \mathrm{A} \tag{6.11}$$

This is matched by a drift current driven by the electric field, caused by the built in voltage acting over the depletion region width. The probable current rating for such a diode in forward bias is less than 1 A, so we will approximate $J_h \simeq 0$ *even* in forward bias. Equation (6.10) may be written:

$$q p \mu_h \mathscr{E} - q D_h \frac{dp}{dx} \simeq 0 \tag{6.12}$$

Using the Einstein relation and $\mathscr{E} = -dV/dx$ this becomes:

$$\frac{dV}{dx} = -\frac{kT}{q} \frac{1}{p} \frac{dp}{dx} \tag{6.13}$$

The next step is an important development. Equation (6.13) is integrated with respect to position x, taking as limits the positions $-d_p$ and $+d_n$ of the depletion region edges measured from the metallurgical junction. The depletion region width is dealt with later, but the values of d_n and d_p do not matter now so long as the integration starts and finishes in the flat portions of the junction band diagram away from the depletion region (Figure 6.6).

$$\int_{-d_p}^{+d_n} \frac{dV}{dx}\, dx = \int_{-d_p}^{+d_n} -\frac{kT}{q}\frac{1}{p}\frac{dp}{dx}\, dx \tag{6.14}$$

Cancelling the dx terms within each integral is equivalent to changing the variables of integration to V and p. The limits must also be changed to the values of V and p at $x = +d_n$ and $x = -d_p$:

$$\int_{V(-d_p)}^{V(d_n)} dV = \int_{p(-d_p)}^{p(d_n)} -\frac{kT}{q}\frac{dp}{p} \tag{6.15}$$

Performing the integration:

$$V(d_n) - V(-d_p) = -\frac{kT}{q}\ln p(d_n)/p(-d_p) \tag{6.16}$$

The lefthand side is simply the total potential drop across the depletion region, V_t. $p(d_n)$ and $p(-d_p)$ are the hole number densities at either edge of the depletion region. The majority carrier densities are almost unchanged by bias, so $p(-d_p) \simeq \bar{p}_p$. The minority carrier density at the junction edge, $p(d_n)$, is not known already and is denoted as p_{n0}. Hence:

$$p_{n0} = \bar{p}_p \exp\frac{-qV_t}{kT} \tag{6.17}$$

which is the Boltzmann expression once again, although there is no equilibrium. Using equations (6.1) and (6.9) this may also be written:

$$p_{n0} = \bar{p}_p \exp\frac{q(V_A - V_B)}{kT} = \bar{p}_n \exp\frac{qV_A}{kT} \tag{6.18}$$

The effectiveness of the forward bias potential is brought out here. Lowering the junction barrier by V_A raises the minority carrier density at the junction edge p_{n0} by a factor of $\exp(qV_A/kT)$ above the equilibrium quantity $\bar{p}_n$. Since $kT/q = 0.026$ electron-volts at room temperature, every increase of V_A by this small amount multiplies the hole number density at the edge of the n-side by 2.7. Again a similar equation may be derived for electrons.

6.4 The diode law

The diode I/V characteristic will be derived quantitatively in this section, using the results of sections 6.2 and 6.3. The key is to realize that the hole current at the edge of the n-type region at the junction is almost entirely a *diffusion* current. Similarly the electron current injected into the edge of the p-type side is a diffusion current. This is because, for low level injection, the bands are virtually flat outside the depletion region. The applied potential is dropped almost entirely across the central zone, leaving only very small electric fields in the bulk of the diode to drive drift current.

This means that the problem is identical to one already solved: the spatial decay of minority carriers. Forward bias raises the minority carrier densities at the edges of the depletion region, according to equation (6.18). Assuming the diode is sufficiently wide, the minority carriers will have all recombined several diffusion lengths away from the junction. Measuring x from the edge of the depletion region:

$$\hat{p}_n = \hat{p}_{n0} \exp -x/L_h \tag{6.19}$$

as shown in section 5.3. The excess minority carrier density is given by:

$$\hat{p}_n = p_n - \bar{p}_n \tag{6.20}$$

which, according to equation (6.18), takes the value:

$$\hat{p}_{n0} = \bar{p}_n \exp \frac{(qV_A)}{kT} - \bar{p}_n$$

$$= \bar{p}_n\left(\exp\left(\frac{qV_A}{kT}\right) - 1\right) \tag{6.21}$$

at $x = 0$. The hole diffusion current can be found from the slope of $\hat{p}_n$ at $x = 0$.

$$J_{hdiff} = -qD_h \left.\frac{d\hat{p}}{dx}\right|_{x=0}$$

$$= \frac{qD_h}{L_h} \bar{p}_n\left(\exp\left(\frac{qV_A}{kT}\right) - 1\right) \mathrm{A\,m^{-2}} \tag{6.22}$$

Adding this to a similar expression for the electron diffusion current at the p-side edge of the depletion region, we obtain the total current density in the diode, as a function of the applied voltage V_A:

$$J = \left[\frac{qD_h\bar{p}_n}{L_h} + \frac{qD_e\bar{n}_p}{L_e}\right]\left(\exp\left(\frac{qV_A}{kT}\right) - 1\right) \mathrm{A\,m^{-2}} \tag{6.23}$$

This can be expressed in terms of the doping on each side, because $\bar{p}_n = n_i^2/\bar{n}_n = n_i^2/N_D$, and $\bar{n}_p = n_i^2/\bar{p}_p = n_i^2/N_A$:

$$J = qn_i^2\left[\frac{D_h}{L_hN_D} + \frac{D_e}{L_eN_A}\right]\left(\exp\left(\frac{qV_A}{kT}\right) - 1\right) \mathrm{A\,m^{-2}} \tag{6.24}$$

Multiplying this by the diode cross-sectional area A gives the total diode current, I:

$$I = I_0\left(\exp\left(\frac{qV_A}{kT}\right) - 1\right) \mathrm{A} \tag{6.25}$$

where I_0 is the constant terms of equation (6.24), multiplied by A. This is the I/V characteristic for a wide *pn* junction diode, and should be learnt. Note that the forward bias current is positive, and consists of holes ($+q$) moving in the $+x$ direction and electrons ($-q$) moving in the $-x$ direction, in keeping with the current convention given in Chapter 1. It is plotted in Figure 6.7, which shows the rectifying behaviour clearly. Three regimes are of interest:

1. $V_A \gg kT/q$. In forward bias the exponential term dominates:

$$I \simeq I_0 \exp\frac{qV_A}{kT} \text{ A}$$ (6.26)

which accounts for the steep rise in current.

2. $V_A = 0$. Putting this in equation (6.25) gives $I = 0$, so the equilibrium condition is satisfied.

3. $V_A \ll -kT/q$. In reverse bias the exponential term rapidly dies away, leaving:

$$I \simeq I_0 \text{ A}$$ (6.27)

I_0 is the reverse leakage current or the reverse saturation current which is constant according to this model. Its value for low current circuit diodes is usually in the range $10^{-9} - 10^{-12}$ A. This is discussed further in the section on practical diode behaviour.

The *pn* junction diode characteristic may be compared with those of the ideal diode (Figure 6.8a) and the semi-ideal diode (Figure 6.8b). In many circuit applications, it is sufficient to say that a diode is on if it has a forward bias above the semi-ideal threshold, and will pass unlimited current. Otherwise it is off and blocks current flow. The semi-ideal threshold voltage is typically 0.7 V in silicon and 0.4 V in germanium. This sometimes causes confusion because these figures are about the same size as the built in voltage V_B. This is a coincidence, and the two quantities have quite different meanings.

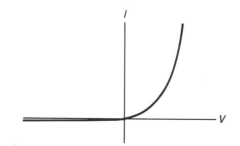

Figure 6.7 The *I/V* characteristic of an ideal *pn* junction diode

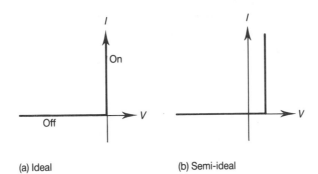

(a) Ideal   (b) Semi-ideal

Figure 6.8 Ideal and semi-ideal diode characteristics

6.5 Drift and diffusion currents in the diode

At each edge of the depletion region, the whole p-side electron current and n-side hole current is a diffusion current. Lowering the junction potential barrier injects carriers into regions where they are minority carriers. They diffuse away from the junction and recombine with majority carriers. Some have likened this to the holes and electrons getting married, but since recombination results in mutual annihilation perhaps the metaphor of the invading minority dying heroically in battle is better. As the minority carrier number density decays and $\mathrm{d}\hat{p}_n/\mathrm{d}x$ falls the diffusion current they carry also declines. If the excess hole density is given by equation (6.19), the diffusion current they carry is:

$$J_{hdiff} = -qD_h \frac{\mathrm{d}\hat{p}_n}{\mathrm{d}x} = \frac{qD_h}{L_h} \hat{p}_{n0} \exp\frac{-x}{L_h} \, \mathrm{A\,m^{-2}} \tag{6.28}$$

Although this is the full injected hole current at $x = 0$, it dies away rapidly. Does this mean that the current in the diode varies? Kirchoff's current law implies not: the current which enters one end of the diode must also flow out of the other end, and flow through all points in between. The decaying diffusion current must be made up by an increasing drift current to make a constant total current.

The drift current depends on the number of mobile charges, their mobility and on the size of the electric field. An electric field away from the depletion region means that some of the applied voltage is dropped across the bulk of the p- and n-regions, rather than the central depletion region. For example, to drive a drift current of 0.1 A through a 1 mm^2 cross-section silicon p-type region doped 10^{24} m^{-3}, $L = 0.5$ mm requires a voltage of $IL/(\mu_h q N_A A)$ across it, which is 6.5 mV. If I_0 is 10^{-10} A the diode law for forward bias (equation (6.26)) gives the applied voltage to permit this current to cross the junction as:

$$V_A \simeq \frac{kT}{q}\ln\frac{I}{I_0} = 0.54 \text{ V} \tag{6.29}$$

With these numbers the error in applied voltage caused by neglecting the bulk voltage drops is 1–2%, with a higher error in the current figure.

Bearing this approximation in mind, the carrier number densities, drift and diffusion currents in forward bias can be represented as in Figure 6.9. Note that a logarithmic scale is used for the number densities, to show both minority and majority carriers. On this scale the exponential decay in injected minority carriers is a straight line, ending at the equilibrium value.

Minority carrier injection ceases in reverse bias. The leakage current which flows consists of minority carriers near the junction crossing over and lowering their energy. This means that the depletion region edges act as sinks for minority carriers, lowering the number density almost to zero. This is shown by putting $V_A \ll 0$ in equation (6.21):

$$\hat{p}_{n0} \simeq -\bar{p}_n \tag{6.30}$$

Hence at the edge of the *n*-side:

$$p_{n0} = \bar{p}_n + \hat{p}_{n0} \simeq 0 \tag{6.31}$$

This depression in carrier density leads to the diffusion of minority carriers towards the junction. Once across they become majority carriers and a very small part of the applied voltage serves to drive a drift current to the contacts. The number densities and currents in a reverse bias diode are shown in Figure 6.10.

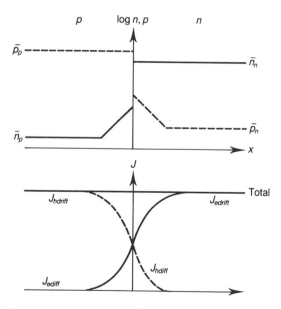

Figure 6.9 Carrier and current distributions in a forward biased *pn* junction

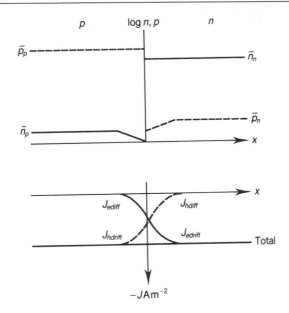

Figure 6.10 Carrier and current distributions in a reverse biased *pn* junction

These figures and the corresponding band diagrams are the most important parts of this chapter. The student who can claim to follow them is well on the way to grasping how a *pn* junction diode works.

6.6 The depletion region

This term was introduced in subsection 4.3.3, where it was defined as a zone void of mobile charges. This means it possesses a net charge/volume equivalent to that of the ionized donors or acceptors present. The band diagram in Figure 6.2b for the *pn* junction in equilibrium clarifies this. Electrons on the *n*-side are contained by the potential barrier, which becomes higher the further they attempt to penetrate the central region. Some will be able to travel a short way in before sliding back, so the charge distribution of Figure 6.2a should be modified as in Figure 6.11. At the outer edge some of the positive donor ion charge is neutralized by mobile electrons. The depletion approximation is to say that this penetration is negligible, so that the charge distribution is purely due to ionized dopants.

The size of the depletion region is important in two respects. First, it sets a limit on the size of the *pn* junction diode and second, it may determine the breakdown voltage of the diode in reverse bias. The potential distribution and band diagram for a single charged region has been given previously. Here it is

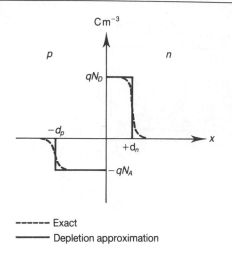

Figure 6.11 The charge distribution at the junction

extended to two charged regions, one with charge density $-qN_A$ coulombs m^{-3} at $-d_p < x < 0$, the other on the *n*-side with a charge density $+qN_D$ at $0 < x < +d_n$ (Figure 6.11).

Poisson's equation (equation 1.9) must be solved in each region separately:

1. *p*-side depletion region:

$$\frac{d^2V}{dx^2} = \frac{qN_A}{\epsilon} \tag{6.32}$$

Integrating twice:

$$\frac{dV}{dx} = \frac{qN_A}{\epsilon} x + C_1 \tag{6.33}$$

$$V = \frac{1}{2} \frac{qN_A}{\epsilon} x^2 + C_1 x + C_2 \tag{6.34}$$

where C_1 and C_2 are constants of integration.

2. *n*-side depletion region:

$$\frac{d^2V}{dx^2} = \frac{-qN_D}{\epsilon} \tag{6.35}$$

Hence:

$$\frac{dV}{dx} = \frac{-qN_D}{\epsilon} x + C_3 \tag{6.36}$$

$$V = -\frac{1}{2}\frac{qN_D}{\epsilon}x^2 + C_3x + C_4 \qquad (6.37)$$

where C_3 and C_4 are unknown constants.

In addition, the widths of the charged regions, d_p and d_n are not known. The following conditions are used to find the six unknown quantities:

1. *Charge neutrality*: the diffusion of charges which set up the junction did not create or destroy charge, so the diode which started out neutral overall remains neutral. For uniform doping levels on each side:

$$qN_Ad_p = qN_Dd_n \qquad (6.38)$$

2. *Continuity of potential*: the metallurgical junction at $x = 0$ has the same potential approached from either side. Further, the zero level of potential is arbitrary, so can be chosen as the value occurring at $x = 0$ for simplicity. Then equations (6.34) and (6.37) provide:

$$C_2 = C_4 = 0 \qquad (6.39)$$

3. *Continuity of field*: the electric field $(-dV/dx)$ is the same at $x = 0$ approached from either side, because the dielectric constant does not change there. Putting $x = 0$ in equations (6.33) and (6.36) gives:

$$C_1 = C_3 \qquad (6.40)$$

4. The depletion approximation provides that the electric field falls to zero at the edges of the depletion region. Hence (dV/dx) is zero at $x = -d_p$ and $x = +d_n$:

$$\frac{-qN_A}{\epsilon}d_p + C_1 = 0 \qquad (6.41)$$

and:

$$\frac{-qN_D}{\epsilon}d_n + C_3 = 0 \qquad (6.42)$$

One more condition is required, because equations (6.41) and (6.42) together are equivalent to equations (6.38) and (6.40).

5. *Total junction potential V_t*: this parameter is known independently, through equations (6.1) and 6.7) and equals the difference in potential between the depletion region edges:

$$V_t = V(d_n) - V(-d_p)$$
$$= \left(-\frac{1}{2}\frac{qN_D}{\epsilon}d_n^2 + C_3d_n\right) - \left(\frac{1}{2}\frac{qN_A}{\epsilon}d_p^2 - C_1d_p\right) \qquad (6.43)$$

There are now enough conditions to solve for the unknowns d_p and d_n in terms of the doping and V_t:

$$d_p = \left(\frac{2\epsilon V_t}{q} \frac{N_D}{N_A(N_A + N_D)} \right)^{1/2} \text{m} \tag{6.44}$$

$$d_n = \left(\frac{2\epsilon V_t}{q} \frac{N_A}{N_D(N_A + N_D)} \right)^{1/2} \text{m} \tag{6.45}$$

The total depletion region width d is the sum of the two:

$$d = d_n + d_p \tag{6.46}$$

Another important result is the maximum electric field $\mathscr{E}_{max}$ within the junction, which may determine the breakdown voltage. Since C_1 and C_3 are both positive, it is easy to see from equations (6.33) and (6.36) that $\mathscr{E}_{max}$ occurs at the metallurgical junction, $x = 0$:

$$\mathscr{E}_{max} = \left(\frac{2qV_t N_A N_D}{\epsilon(N_A + N_D)} \right)^{1/2} \text{V m}^{-1} \tag{6.47}$$

These complex expressions have simpler forms for single-sided abrupt junctions. For example, in a p^+n device where $N_A \gg N_D$:

$$d_p \simeq \left(\frac{2\epsilon V_t N_D}{q N_A{}^2} \right)^{1/2} \tag{6.48}$$

$$d_n \simeq \left(\frac{2\epsilon V_t}{q N_D} \right)^{1/2} \tag{6.49}$$

$$d \simeq d_n \tag{6.50}$$

$$\mathscr{E}_{max} \simeq \left(\frac{2qV_t N_D}{\epsilon} \right)^{1/2} \tag{6.51}$$

These results can be summarized:

1.   Most of the depletion region occurs on the lightly doped side. This is also clear from the charge balance condition, equation (6.38).
2.   The maximum electric field in a single-sided junction is determined by the lighter doping level.
3.   The depletion region width and peak field in reverse bias vary as the square root of the applied voltage (assuming $|V_A| \gg V_B$).

6.7 Reverse bias breakdown

The ability of the *pn* junction to resist current flow in reverse bias is not unlimited. At some point it will break down and permit a large current to flow. There are three principal breakdown mechanisms, illustrated in Figure 6.12:

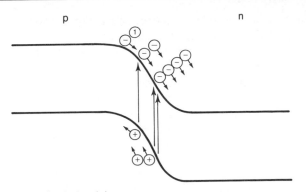

(a) Avalanche breakdown

Note:
An electron ① excites another into the conduction band by collision. These excite more electrons in turn. The extra holes can also create more pairs at position ① when they travel towards the *p*-region. This gives a runaway current.

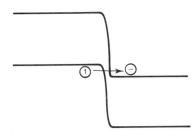

(b) Tunnel breakdown.

Note.
An electron in the *p*-side valence band at position ① can lower its energy if it can tunnel to the *n*-side conduction band. This requires strong reverse bias and a narrow depletion region (high doping).

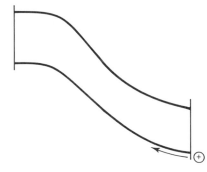

(c) Punch through breakdown.

Note.
The depletion region reaches one of the contacts, which can supply minority carries easily. These are swept across the diode by the high depletion region field.

Figure 6.12 Three principal breakdown mechanisms

1. *Avalanche breakdown*: if an electron can acquire enough energy from the electric field, it can knock another electron into the conduction band. This in turn can generate further electron–hole pairs, so the current snowballs. The effect is intensified by the fact that the holes so formed travel back in the opposite direction and can also cause extra ionization in collisions. Avalanche breakdown will occur if the maximum field within the junction $\mathscr{E}_{max}$ exceeds a critical value for that material, $\mathscr{E}_{aval}$. Strictly the field must also exist over a long enough distance, but this is usually satisfied automatically in a depletion region.

2. *Tunnel breakdown*: this applies especially to heavily doped diodes with thin depletion regions. Tunnelling is a quantum-mechanical phenomenon, and relies on the wave-like behaviour of electrons. Faced with an energy barrier of width d and height E, which in this case is the band gap, an electron has a probability of penetrating it. Even though it does not have sufficient energy to hop over the barrier, it may tunnel through. The probability of tunnelling is greater for smaller values of E and d, favouring thin depletion regions.

Zener breakdown is a mixture of these two and is characterized by occurring at a well-defined reverse bias potential. This makes it useful as a voltage reference, as will be discussed when describing practical varieties of diode.

3. *Punchthrough breakdown*: if the depletion region becomes large enough to reach either one of the contacts to the semiconductor, or to a forward biased junction, then a large current will flow. This is because the contacts are assumed to be low resistance providers of minority carriers, and the depletion region is a zone of high electric field. When the two meet the minority carrier current increases dramatically.

Breakdown is usually to be avoided, unless it serves a useful function, such as voltage regulation, or protecting more sensitive devices by providing a safe dissipation path for energetic pulses. It need not be destructive, provided the current which flows is limited by the external circuit impedance. If an excessive current flows the local heating will destroy a device far faster than an alarmed observer will be able to turn it off.

EXAMPLE 1

■ Draw the band diagram for a p^+n junction in forward bias, and the distribution of holes and electrons through the device.

□ Start by drawing the equilibrium band diagram, with the Fermi level constant (Figure 6.13a). The depletion region is nearly all on the *n*-type side, because it is doped more lightly (equation (6.38)). Forward bias lowers the levels on the p-side by the applied potential V_A, and separates the Fermi levels by the same amount (Figure 6.13b). The depletion region shrinks slightly, according to equation (6.48), because the total barrier potential V_t is reduced by V_A. There is a small slope on the bands outside the depletion region because of the current flow. The slope tends to drive majority carriers towards the junction.

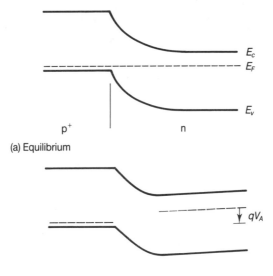

Figure 6.13 p^+n junction in equilibrium and forward bias

The carrier distributions are built up as shown in Figure 6.14. Well away from the junction the carrier levels are flat, close to their equilibrium values. In a p^+n junction the majority carrier density is higher and the minority density is lower on the p-side. On a logarithmic scale, the sum of $\bar{p}_n$ and $\bar{n}_p$ equals the sum of $\bar{n}_n$ and $\bar{p}_n$, because $pn = n_i^2$ in each material in equilibrium. Close to the junction the minority carrier densities are raised by injection. The levels are raised by a factor $\exp(qV_A/kT)$ on both sides, which is equivalent to the same vertical increase on a log scale. They decay exponentially to their bulk values away from the junction, which is a straight line on a log scale. The steeper slope belongs to the species with the shorter minority carrier diffusion length, normally holes.

The small imbalance in total charge in the minority carrier decay regions causes the electric field which drives the drift currents towards the junction.

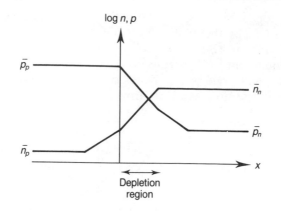

Figure 6.14 Carrier distribution in a forward biased p^+n diode

EXAMPLE 2

■   Show that the form of the diode law is obeyed in a short diode, in which the *n*- and *p*-side widths W_n and W_p are much shorter than the minority carrier diffusion lengths L_h and L_e respectively. Find an expression for the reverse leakage current in a short diode. Assume that the excess minority densities are zero at the contacts, and the depletion region does not approach the contacts.

☐   The number density of excess minority carriers at the depletion zone edges do not depend on the width of the near-neutral regions, provided they exist at all. Equation (6.21) still applies:

$$\hat{p}_{n0} = \bar{p}_n\left(\exp\left(\frac{qV_A}{kT}\right) - 1\right) \tag{6.21}$$

The exponential decay of excess carriers is no longer valid, because that requires $W_n, W_p \gg L_h, L_e$. The diffusion equation (equation (5.34) for holes) must be solved anew with fresh boundary conditions:

$$\frac{\mathrm{d}^2\hat{p}_n}{\mathrm{d}x^2} - \frac{\hat{p}_n}{L_h^2} = 0 \tag{6.52}$$

$$\hat{p}_n = \hat{p}_{n0}, \ x = 0$$

$$\hat{p}_n = 0, \ x = W_n, \text{ the contact edge}$$

The exact solution is complex, but a good approximation simplifies it a lot. If $L_h \gg W_n$ this is equivalent to saying that very little recombination takes place in the *n*-side. Almost all the excess carriers reach the contacts and recombine there. This means that the recombination term $(\hat{p}_n/L_h{}^2)$ can be neglected. (This statement can be checked by considering its origin in section 5.3.) The diffusion equation becomes:

$$\frac{d^2\hat{p}_n}{dx^2} \simeq 0 \tag{6.53}$$

Integrating directly:

$$\frac{d\hat{p}_n}{dx} = C_1 \tag{6.54}$$

$$\hat{p}_n = C_1 x + C_2 \tag{6.55}$$

The constants of integration C_1 and C_2 are found using the boundary conditions:

$$\hat{p}_n = \frac{-\hat{p}_{n0}}{W_n} x + \hat{p}_{n0} \tag{6.56}$$

This is plotted in Figure 6.15. The excess hole density decays linearly in the absence of recombination. (The full solution would make the line curve downwards slightly.) The diffusion current associated with the injected minority

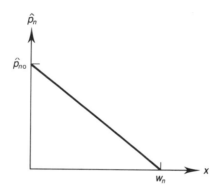

Figure 6.15 Minority carrier decay in a short diode

carriers does not die away in this case but is constant, because the line has a constant slope. Its value is:

$$J_{hdiff} = -qD_h \frac{d\hat{p}_n}{dx} = qD_h\hat{p}_{n0}/W_n \ \text{A m}^{-2} \tag{6.57}$$

with a similar expression for electrons. The total current density in the diode is the sum of the hole and electron diffusion currents:

$$J = q\left[\frac{D_h \bar{P}_n}{W_n} + \frac{D_e \bar{n}_p}{W_p}\right]\left(\exp\left(\frac{qV_A}{kT}\right) - 1\right) \text{A m}^{-2} \tag{6.58}$$

which may be compared with equations (6.23) for the wide diode. The short diode has an identical I/V characteristic, with W_n and W_p replacing L_h and L_e in the expression for the reverse leakage current.

EXAMPLE 3

■   An abrupt *pn* diode is doped $N_A = 10^{22}$ and $N_D = 10^{24}$ m^{-3} on either side. Find the built in voltage and the depletion region width in equilibrium at 300 K. Find whether avalanche or punchthrough breakdown occurs first, if $W_p = 5\ \mu$m, $W_n = 100\ \mu$m and the avalanche field is 2×10^7 V m^{-1}. Take $n_i = 10^{16}$ m^{-3}, $\epsilon_r = 4$.

☐   The built in voltage is found using equation (6.7):

$$V_B = \frac{kT}{q} \ln \frac{N_A N_D}{n^2_i} = 0.83 \text{ V}$$

Since $N_D \gg N_A$, the diode is an n^+p single-sided abrupt junction and most of the depletion region is on the p-type side. Using this approximation equation (6.44) becomes:

$$d_p \simeq \left(\frac{2\epsilon V_t}{qN_A}\right)^{1/2}$$

In equilibrium $V_t = V_B$. The permittivity ϵ is the product of the relative permittivity ϵ_r and the value for free space, $\epsilon_0 = 8.85 \times 10^{-12}$; hence the depletion region width is:

$$d \simeq d_p = 0.19\ \mu\text{m}$$

The maximum electric field in the junction is found by putting $N_D \gg N_A$ in equation (6.47):

$$\mathscr{E}_{max} = \left(\frac{2qV_t N_A}{\epsilon}\right)^{1/2}$$

In reverse bias both this and the p-side depletion region width increase. The device will fail either when $\mathscr{E}_{max}$ exceeds the avalanche threshold value or when d_p exceeds the p-side depletion width. Recasting these equations in terms of d_p:

$$V_t = \frac{qN_A d_p^{\ 2}}{2\epsilon}$$

and

$$V_t = \frac{\epsilon}{2qN_A} \mathscr{E}_{max}{}^2$$

in terms of $\mathscr{E}_{max}$. If we put $d_p = W_p$ and $\mathscr{E}_{max} = \mathscr{E}_{aval}$ in these:

$V_t = 565$ V at punchthrough

$V_t = 4.4$ V at avalanche

The figure for avalanche is far lower, so as the barrier height is increased in reverse bias it fails by avalanche before it fails by punchthrough.

Summary

Main symbols introduced

Symbol	Unit	Description
d	m	total depletion region width
d_n, d_p	m	width of the depletion zone on the n-side, p-side
$\mathscr{E}_{aval}$	$V\,m^{-1}$	avalanche breakdown field for the material
$\mathscr{E}_{max}$	$V\,m^{-1}$	maximum electric field in the device
I_0	A	reverse leakage current
n_n, n_p	m^{-3}	electron number density on the n-side, p-side
p_n, p_p	m^{-3}	hole number density on the n-side, p-side
V_A	V	potential applied to the diode
V_B	V	built in junction potential
V_t	V	total barrier potential
W_n, W_p	m	width of the diode n-side, p-side
V_n, V_p	V	local potential on the n-side, p-side

Main equations

$$V_t = V_B - V_A \text{ V} \quad (V_A > 0 \text{ in forward bias, } V_B > 0) \tag{6.1}$$

$$V_B = \frac{kT}{q} \ln \frac{N_A N_D}{n_i{}^2} \text{ V} \tag{6.7}$$

$$p_{n0} = \bar{p}_n \exp \frac{qV_A}{kT} \quad \text{(at the depletion region edge)} \tag{6.18}$$

$$I = I_0\left(\exp\left(\frac{qV_A}{kT}\right) - 1\right) \quad \text{(diode law)} \tag{6.25}$$

$$I_0 = qn_i^2 A\left(\frac{D_h}{L_h N_D} + \frac{D_e}{L_e N_A}\right) \quad \text{(wide diode)} \tag{from 6.24}$$

$$I_0 = qn_i^2 A\left(\frac{D_h}{W_n N_D} + \frac{D_e}{W_p N_A}\right) \quad \text{(short diode)} \tag{from 6.58}$$

$$N_A d_p = N_D d_n \quad \text{(neutrality condition)} \tag{6.38}$$

$$d_p = \left(\frac{2\epsilon V_t}{q}\frac{N_D}{N_A(N_A + N_D)}\right)^{1/2} \quad \text{(for an abrupt junction)} \tag{6.44}$$

$$d_n = \left(\frac{2\epsilon V_t}{q}\frac{N_A}{N_D(N_A + N_D)}\right)^{1/2} \quad \text{(for an abrupt junction)} \tag{6.45}$$

$$E_{max} = \left(\frac{2qV_t}{\epsilon}\frac{N_A N_D}{(N_A + N_D)}\right)^{1/2} \quad \text{(for an abrupt junction)} \tag{6.47}$$

Exercises

1. Sketch the band diagrams for a pn^+ junction in equilibrium, forward bias and reverse bias.

2. Draw the electron and hole number density distributions and the drift and diffusion current variation with position through a wide pn^+ abrupt junction in forward and reverse bias.

3. Calculate the reverse leakage current in a *pn* diode for which $N_A = N_D = 10^{23}$ m^{-3}, $n_i = 10^{16}$ m^{-3}, $W_n = W_p = 6$ μm, $\tau_e = \tau_h = 3$ μs, $\mu_e = 0.1$ m^2 V^{-1}s^{-1}, $\mu_h = 0.05$ m^2 V^{-1}s^{-1}, and the cross-sectional area is 0.8 mm^2. Find the currents which flow when $V_A = \pm 2$ V. Take $T = 300$ K.

4. Does the built in voltage vary linearly with temperature? Could the variation be used to measure temperature?

5. Find the ratio of electron and hole injection currents in a wide abrupt *pn* diode with $n_i = 10^{16}$ m^{-3}, $N_A = 10^{20}$ m^{-3}, $N_D = 10^{23}$ m^{-3}, $\mu_e = 0.14$ m^2 V^{-1}s^{-1}, $\mu_h = 0.048$ m^2 V^{-1}s^{-1}, $\tau_e = \tau_h = 20$ μs, $T = 300$ K. Find the reverse leakage current if the cross-sectional area is 1 mm^2.

6. Solve Poisson's equation for the *n*-side depletion region of a p^+n junction to verify the results of equations (6.49) and (6.51).

7. A single-sided abrupt n^+p junction has the following characteristics at 300 K:
$\epsilon_r = 11.8$, $N_D = 10^{25}$ m^{-3}, $N_A = 10^{21}$, $n_i = 1.5 \times 10^{16}$ m^{-3}, $\mathscr{E}_{aval} = 3 \times 10^7$ V m^{-1},
$W_p = 2\ \mu$m. Find the reverse breakdown mechanism and the breakdown voltage.
How could the breakdown voltage be increased, and how would I_0 and the on
resistance be affected by the change?

8. The temperature of the diode in (3) is raised to 400 K. Find the reverse saturation
current if the bandgap is 1.1 eV, and the mobilities are constant. (The variation of
I_0 with temperature is a serious limitation in circuit design.)

9. Choose appropriate doping levels for an abrupt pn^+ junction diode with the
following constraints: maximum breakdown voltage; minimum 'on' resistance when
forward biased; $W_p = 0.2$ mm. Take $n_i = 10^{16}$ m^{-3}, $\epsilon_r = 12$, $\mathscr{E}_{aval} = 3 \times 10^7$ V m^{-1}.

10. The internal resistance of a wide n^+p diode is approximately that of the more
resistive side. If $N_D = 10^{25}$ m^{-3}, $N_A = 10^{23}$ m^{-3}, $n_i = 1.5 \times 10^{16}$ m^{-3}, $T = 300$ K,
$\mu_e = 0.1$, $\mu_n = 0.05$ m^2 V^{-1} s^{-1}, $L_e = L_h = 10\ \mu$m, $W_p = 0.1$ mm and the
cross-sectional area is 0.5 mm^2, find the applied voltage which will give a forward
current of 1 A. Take into account the voltage drop across the p-side.

7 Ac effects and real diodes

In Chapter 6 we discussed the response of a *pn* junction to steady applied voltages and saw that the current could be regulated by varying the height of the junction barrier. In forward bias it is lowered to permit minority carrier injection, but in reverse bias it is increased and only a small leakage current of minority carriers flows. In this section we are concerned with the behaviour of the diode in response to time-varying or alternating (ac) voltages. Two types of ac voltage are of interest (Figure 7.1):

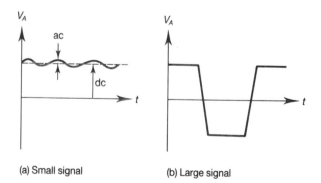

(a) Small signal   (b) Large signal

Figure 7.1 Two types of ac voltage

1. Small-signal ac, where the time-varying component of the applied voltage is much smaller than the steady dc component.
2. Large-signal ac, or switching waveforms, where the diode is driven between forward and reverse bias.

In the small-signal case it is possible to derive an equivalent circuit for the diode, which may be used to model the junction's behaviour when it forms part of a larger small-signal circuit. The switching or transient case is not amenable to simple analysis but a qualitative picture may be formed which explains the observed behaviour quite well.

7.1 Small-signal response

7.1.1 Slope resistance, r_e

First we consider the small-signal or differential resistance of the diode. This is the resistance seen by the *ac part* of a combined ac + dc applied potential. The distinction is brought out in Figure 7.2. A diode forward biased to point P passes a dc current such that the dc resistance is $V_A/I = OQ/QP$. This takes on different values as the basis is altered, so we say that the diode is a non-linear dc resistance. (This is a bit of a nonsense, because the term resistance might seem to imply that Ohm's law is obeyed. A diode clearly does not, but the ratio volts/amps is still measured in ohms.)

If the forward bias is altered from its value at P by a small amount ΔV_A, the current increment is ΔI. The ratio of these quantities is the *differential resistance* r_e:

$$r_e = \frac{\Delta V_A}{\Delta I} = \frac{PR}{SR}\ \Omega \qquad (7.1)$$

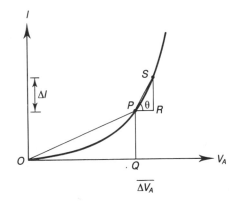

Figure 7.2 Differential resistance, $r_e = \Delta V_A/\Delta I = PR/SR$

This means that if the diode has a steady dc bias and a small ac signal is then applied, the current increase is $\Delta V_A/r_e$, where both ΔV_A and r_e are small-signal quantities. It is essential to keep ac and dc calculations in watertight compartments. The diode does *not* possess a resistance r_e to steady applied voltages: it is a diode and obeys the diode law.

The differential resistance, r_e, is not itself a fixed quantity because the diode I/V characteristic does not possess a constant slope. This is seen by differentiating the diode law:

$$I = I_0\left(\exp\left(\frac{qV_A}{kT}\right) - 1\right)$$

(7.2)

Therefore:

$$\frac{1}{r_e} = \frac{dI}{dV_A} = \frac{q}{kT} I_0 \exp\frac{qV_A}{kT} \simeq \frac{qI}{kT}$$

(7.3)

In forward bias the unity term in the diode law may be neglected, so:

$$r_e \simeq \frac{kT}{qI}$$

(7.4)

At low current the characteristic is nearly flat and I is changed very little by an increase in V_A. This corresponds to a large value of r_e. When I is larger r_e falls so that the same increase in V_A produces a far greater swing in current. The approximation $I \gg I_0$ fails in reverse bias and r_e becomes very large. An increase in reverse bias produces virtually no increase in current because it is nearly constant at $-I_0$.

7.1.2 Junction capacitance C_j

There are two sources of capacitative impedance in the *pn* diode. The first is called junction capacitance and is due to the charge stored in the depletion region. Figure 7.3 shows the response of the depletion region to a change in applied voltage: it expands or contracts as the junction is moved away from or towards forward bias. The size of the charge is the same on each side of the junction:

$$Q_j = qN_D d_n = qN_A d_p$$

(7.5)

Using equation (6.46) for the depletion region width this becomes:

$$Q_j = N_A \left(\frac{2\epsilon V_t q N_D}{N_A(N_A + N_D)}\right)^{1/2} \mathrm{C\,m}^{-2}$$

(7.6)

Junction capacitance C_j is a small-signal quantity, so is defined:

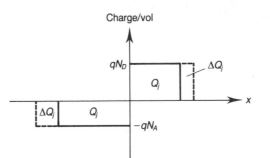

Figure 7.3 Junction capacitance charge

$$C_j = \frac{-\mathrm{d}Q_j}{\mathrm{d}V_A} \; \mathrm{F\,m}^{-2} \tag{7.7}$$

The minus sign is because an increase in Q_j is caused by a decrease in V_A. (See equation (7.8) and think about it.) Note that C_j is in farads per square metre and must be multiplied by the cross-sectional area of the junction to give the total circuit capacitance. The total barrier voltage V_t in equation (7.6) is composed of the constant built in voltage and the applied voltage:

$$V_t = V_B - V_A \; \mathrm{V} \tag{7.8}$$

where V_B is positive and V_A is positive in forward bias. Therefore, differentiating equation (7.6) with respect to V_A gives:

$$C_j = \left(\frac{\epsilon q N_D N_A}{2(N_A + N_D)V_t} \right)^{1/2} \mathrm{F\,m}^{-2} \tag{7.9}$$

C_j is proportional to $V_t^{-1/2}$ for an abrupt junction. (Exercise 3 is to show that C_j varies as $V_t^{-1/3}$ in a linear junction.) The junction capacitance is lower at high reverse bias because the extra charges are added at a greater separation as the depletion region expands. This source of capacitance is present in both forward and reverse bias.

7.1.3 Diffusion capacitance C_d

The second mechanism of charge storage in a *pn* diode is the injected minority carrier charge. Forward bias raises the minority number density at the depletion region edge and these diffuse away from the junction and recombine. The spatial decay is exponential for a wide diode and tends to a linear limit for a short diode (Figure 7.4). The total charge stored is the area under the appropriate curve. For the wide diode ($W_n \gg L_h$):

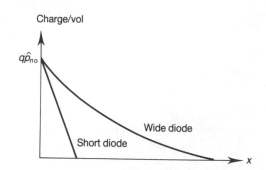

Figure 7.4 Injected minority carrier storage in two diodes

$$Q_d = \int_0^\infty q\hat{p}_{n0} \exp\frac{-x}{L_h}\, dx$$

$$= \left[-q\hat{p}_{n0}\, L_h \exp\frac{-x}{L_h} \right]_0^\infty$$

$$= q\hat{p}_{n0}\, L_h \, \mathrm{C\, m^{-2}} \tag{7.10}$$

For the narrow diode ($W_n \ll L_h$):

$$Q_d = \int_0^{W_n} q\hat{p}_{n0}\left(1 - \frac{x}{W_n}\right) dx$$

$$= q\hat{p}_{n0}\left[x - \frac{x^2}{2W_n} \right]_0^{W_n}$$

$$= \frac{1}{2}\, q\hat{p}_{n0}\, W_n \, \mathrm{C\, m^{-2}} \tag{7.11}$$

The small-signal diffusion capacitance is defined as:

$$C_d = \frac{dQ_d}{dV_A} \, \mathrm{F\, m^{-2}} \tag{7.12}$$

This is positive because an increase in V_A (more forward bias) leads to an increase in $\hat{p}_{n0}$:

$$\hat{p}_{n0} = \bar{p}_n \exp\frac{qV_A}{kT} \, \mathrm{m^{-3}} \tag{7.13}$$

Hence for the wide diode the diffusion capacitance is:

$$C_d = \frac{q^2 L_h}{kT}\, \bar{p}_n \exp\frac{qV_A}{kT} \, \mathrm{F\, m^{-2}} \tag{7.14}$$

This strong non-linear dependence on V_A means that the diffusion capacitance is most effective in forward bias. There is only a small effect in reverse bias,

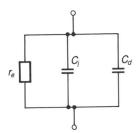

Figure 7.5 Small-signal equivalent circuit of a *pn* junction

because the maximum departure from neutrality outside the depletion region is only $q\bar{p}_n$, caused by the loss of minority carriers.

The slope resistance r_e is in parallel with the two sources of capacitance. For small-signal analysis *only* the *pn* junction diode can be replaced with the equivalent lumped circuit shown in Figure 7.5.

EXAMPLE

■ Find the size of the small-signal equivalent circuit elements for a wide pn^+ diode of cross-sectional area 2 mm², $N_A = 10^{20}$ m⁻³, $N_D = 10^{25}$ m⁻³, $L_e = 20\ \mu$m, at 300 K with 0.4 V forward bias. Take $\mu_e = 0.13$ m² V⁻¹ s⁻¹, $n_i = 10^{16}$ m⁻³, $\epsilon_r = 12$.

☐ The slope resistance is found using the diode law (equation (6.25)) to obtain the current, assuming electron injection dominates:

$$I = A\,\frac{n_i{}^2 kT\mu_e}{L_e N_A}\left(\exp\left(\frac{qV_A}{kT}\right) - 1\right) = 0.291\ \text{mA}$$

The Einstein relation has been used. The approximation equation (7.4) is valid because $I \gg I_0$ with this forward bias. Hence:

$$r_e = \frac{kT}{qI} = 88.6\ \Omega$$

The junction capacitance can be found from equation (7.9) provided the built in voltage is known:

$$V_B = \frac{kT}{q}\ln\frac{N_A N_D}{n_i{}^2} = 0.772\ \text{V}$$

The barrier potential V_t is therefore $0.772 - 0.4 = 0.372$ V. Hence:

$$C_j \simeq \left(\frac{\epsilon q N_A}{2V_t}\right)^{1/2} = 4.78 \times 10^{-5}\ \text{F m}^{-2}$$

With a cross-sectional area of $2\,\mathrm{mm}^2$, this gives a small-signal circuit value of $95.6\,\mathrm{pF}$.

The diffusion capacitance can be found from the p-side equivalent of equation (7.14):

$$C_d = \frac{q^2 L_e \bar{n}_p}{kT} \exp \frac{qV_A}{kT} = 6.69 \times 10^{-4}\,\mathrm{F\,m^{-2}}$$

which gives a total of $1.34\,\mathrm{nF}$ for C_d. In this forward biased junction the small-signal capacitance is dominated by the charge of the injected minority carriers.

7.2 Transient response

When the change in applied potential is not small compared to any steady level, small-signal analysis fails because C_j, C_d and r_e can no longer be regarded as constants over the swing in voltage. The problem becomes more complex because we are talking of charging and discharging capacitances whose values depend on the voltage across them. Insight can be gained by considering a sharp square wave signal of amplitude $\pm V$ driving a diode in series with a fixed resistance R (Figure 7.6). In forward and reverse bias Kirchoff's law for the circuit gives:

$$+V = I_f R + V_{Af} \quad \text{(forward)} \tag{7.15}$$

$$-V = I_r R + V_{Ar} \quad \text{(reverse)} \tag{7.16}$$

V_{Af} and V_{Ar} are the diode voltage drops in forward and reverse bias. Since the junction behaves like a capacitance the voltage across the junction at the *instant* of switching is unaltered. This is confirmed by equation (7.13). If the minority carrier density is unchanged – and it cannot change instantly – then V_A is unaltered. Hence:

$$V_{Af} = V_{Ar} \text{ at the moment of switching} \tag{7.17}$$

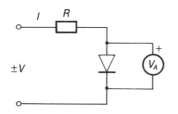

Figure 7.6 Diode switching

This means that V_A is pinned to its initial value in both size and polarity. The implication for the current flow is given by subtracting equations (7.15) and (7.16):

$$2V = R(I_f - I_r)$$

or:

$$I_r = I_f - \frac{2V}{R} \tag{7.18}$$

Since the forward current is of order $+V/R$ if $V \gg V_A$, the reverse current which flows at the moment of transition is about $-V/R$. This has serious consequences for the choice of component in switching circuits, because for a short period after applying reverse bias the diode conducts freely in the reverse direction.

This phenomenon can be explained in terms of the injected minority charge. Consider the transition from forward to reverse bias. The minority carrier distribution must change as indicated in Figure 7.7 from diffusion away from the junction to a small flux towards the junction. The intermediate stages are also shown. Minority carrier injection slows at the moment of transition because majority carriers are being drawn away from the junction by the new potential. There are three ways of dissipating the excess minority carriers:

1.   Recombination with majority carriers.
2.   Diffusion to the contact.
3.   Diffusion back across the junction.

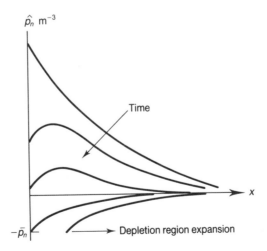

Figure 7.7 Minority carrier distribution when switching from forward to reverse bias

The last route may appear closed because the slope at $t = 0$ serves to diffuse carriers away from the junction. However, particles are moving randomly in all directions. At a distance from the edge carriers are available from both directions so Fick's diffusion law applies. At the depletion region edge itself the cessation of injection means that the random flux back across the junction is not countered. This leads to a rapid fall in the excess number density at the edge, sufficient to change the sign of the slope of $\hat{p}_n(x)$. Diffusion back towards the junction follows.

A detailed analysis shows that this diffusion flux is nearly constant during the initial phase of the switching, so that the reverse current is almost constant. This period is known as the storage phase of duration t_s and can be seen on a current/time plot such as Figure 7.8. It is followed by a decay phase during which the remaining charge decays and the depletion region expands (C_j charges) to its steady reverse bias position.

Commercially available diodes are often quoted for a value of t_{rr}, the reverse recovery time, being the time from the onset of switching to that of the diode attaining 95% of its final voltage. Gold-doped silicon signal diodes have t_{rr} values of a few nanoseconds, while in larger rectifying diodes with a few amps current capability a few microseconds is more typical.

The effect can be readily demonstrated in the laboratory with such devices,

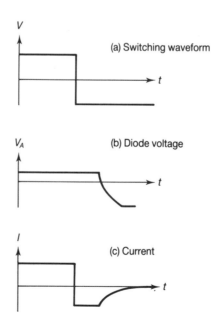

Figure 7.8 Switching response

where the minority carrier lifetime can be measured by plotting a graph of the form:

$$t_s = \tau_h \ln\left(1 + \left|\frac{I_f}{I_r}\right|\right)$$ (7.19)

for various values of the square-wave driving voltage. Larger power rectifiers have an even longer t_{rr} but the effect is minimized in devices which deal with sinusoidal voltages because the stored minority charge is reduced steadily to zero at 50–60 Hz. The diodes nearly follow the diode law under these conditions.

The transition from reverse to forward bias is much more rapid in switching, being limited mainly by the discharge of the depletion region capacitance through the circuit series impedance.

7.3 Real diodes

A perfect diode would have an I/V characteristic like that in Figure 7.9a. It would pass unlimited current with no voltage drop in the forward direction and block all current at any reverse bias. Real diodes are good approximations to semi-ideal or piece-wise linear diodes (Figure 7.9b and c), which pass unlimited or resistance-limited current provided the forward bias tends to exceed a certain threshold or turn on voltage. I/V plots for germanium (band gap 0.67 eV), silicon (1.1 eV), and gallium arsenide (1.4 eV) diodes in forward bias at low temperature are shown in Figure 7.10, assuming $N_A = N_D = 100\ n_i$ in each case. The smaller band gap materials display a lower turn on voltage. This is borne out by the diode law for this case in forward bias:

$$I = Aqn_i \left(\frac{D_h}{100L_h} + \frac{D_e}{100L_e}\right) \exp \frac{qV_A}{kT}$$ (7.20)

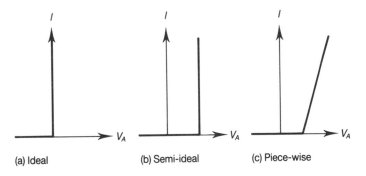

(a) Ideal   (b) Semi-ideal   (c) Piece-wise

Figure 7.9 Ideal, semi-ideal and piece-wise linear diodes

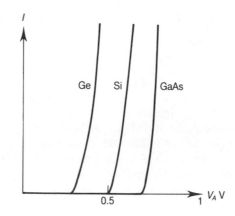

Figure 7.10 Turn on characteristics of Ge, Si and GaAs diodes

Expressing n_i in terms of the band gap and lumping the constants we get:

$$I = C \exp \frac{2qV_A - E_g}{kT} \tag{7.21}$$

The exponential term will be small until $2qV_A$ exceeds E_g by several kT. Figure 7.10 confirms that each diode turns on when V_A is a little more than half the band gap (expressed in eV).

Measured I/V characteristics often depart from the ideal diode law, one reason being the neglect of the generation and loss of electron–hole pairs in the depletion region. Consider first what happens if a pair is created (say by thermal excitation) within the depletion region (Figure 7.11). A high electric field exists there due to the built in potential, which tends to draw the mobile charges apart. The hole is driven to the p-side and the electron to n-side. The band diagram shows this as the hole tending to rise and the electron to fall. The

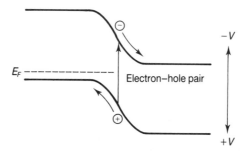

Figure 7.11 Motion of charges generated within the depletion region

currents associated with the two moving charges reinforce each other because the opposite charges are headed in opposite directions.

The current is especially noticeable in reverse bias, when the higher potential barrier excludes majority carriers from the depletion region. This means that no recombination balances the generation. In reverse bias the generation current reinforces the reverse current because of the diffusion of minority carriers back across the junction, so the current becomes:

$$I = -(I_0 + I_{GR0}) \text{ A} \quad \text{(reverse bias)} \tag{7.22}$$

The subscript GR stands for generation–recombination. The size of I_{GR} in reverse bias, I_{GR0}, depends on the volume of the depletion region, Ad, where A is the cross-sectional area and d the depletion region width. If it is assumed that traps at the intrinsic Fermi level E_{Fi} are responsible for the generation, and that the electron and hole capture coefficients are equal, then it can be shown that:

$$I_{GR0} = \frac{qAdn_i}{2\tau} \text{ A} \tag{7.23}$$

where $\tau = \tau_e = \tau_h$, the minority carrier lifetime. The importance of this term becomes clear if we take the ratio of I_{GR0}/I_0, using the results of equation (6.25):

$$\frac{I_{GR0}}{I_0} = \frac{d}{2\tau n_i \left(\dfrac{D_h}{L_h N_D} + \dfrac{D_e}{L_e N_A} \right)} \tag{7.24}$$

For silicon at 300 K with n_i 1.5×10^{16}, let us take $N_A = N_D = 10^{22}$ m^{-3}, $D_e = 3.5 \times 10^{-3}$ m^2 s^{-1}, $D_h = 1.2 \times 10^{-3}$, $\tau_e = \tau_h = 10^{-5}$ s. This gives $L_e = 1.9 \times 10^{-4}$ m, $L_h = 1.1 \times 10^{-4}$, and the depletion region width $d = 2.1 \times 10^{-7}$ m without bias. Hence:

$$\frac{I_{GR0}}{I_0} = 239$$

The generation term is therefore dominant under these conditions, and the actual reverse saturation current will be considerably higher than that predicted by an ideal theory.

In forward bias the same depletion region traps cause some of the majority carriers in transit through the region to recombine. The number density of carriers on each side is still determined by the Boltzmann relation (equation (6.18)), so the recombining carriers constitute an addition to the normal flow. The direction of current flow is opposite to that in the reverse bias generation case because now a hole comes from the p-side and an electron from the n-side, to be lost in the depletion region. It can be shown that in forward bias:

$$I_{GR} = I_{GR0} \exp \frac{qV_A}{2kT} \text{ A} \tag{7.25}$$

Note the factor two in the exponent, which does not appear in the diode law. The total forward bias current is therefore:

$$I = I_0 \exp\frac{qV_A}{kT} + I_{GR0} \exp\frac{qV_A}{2kT} \text{ A} \tag{7.26}$$

which is often approximated as:

$$I = I_0' \exp\frac{qV_A}{nkT} \text{ A} \tag{7.27}$$

where n is the diode ideality factor, and lies between one and two. At high forward bias resistive loss in the neutral region becomes important, so that V_A must be replaced with $V_A - IR$. The resistance R will also depend on the current in these circumstances because the injected minority carriers begin to influence the conductivity. The real diode characteristic is presented in Figure 7.12 and includes the notion of breakdown discussed previously.

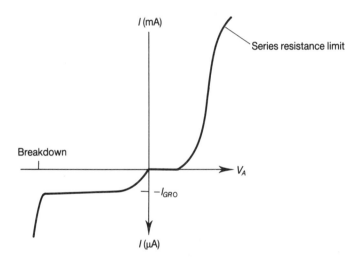

Figure 7.12 Real I/V characteristic (note the expanded scale in reverse bias)

EXAMPLE

■ A *pn* diode has values of 1 pA for I_0 and 1 nA for I_{GR0}. Plot a graph to determine the diode ideality factor and the pre-exponential current I_0' in equation (7.27). Take $kT/q = 0.026$ V.

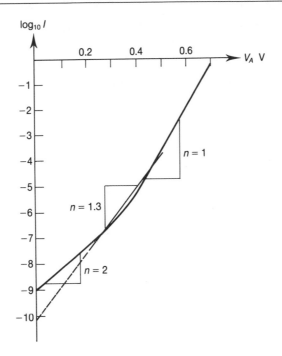

Figure 7.13 Ideality factors in forward bias

☐ The *I/V* characteristic in forward bias is derived from equation (7.26) (Figure 7.13).

V_A	$I_0 \exp qV_A/kT$	$I_{GR0} \exp qV_A/2kT$	I
0.1	4.7×10^{-11}	6.8×10^{-9}	6.9×10^{-9}
0.2	2.2×10^{-9}	4.7×10^{-8}	4.9×10^{-8}
0.3	1.0×10^{-7}	3.2×10^{-7}	4.2×10^{-7}
0.4	4.8×10^{-6}	2.2×10^{-6}	7.0×10^{-6}
0.5	2.2×10^{-4}	1.5×10^{-5}	2.4×10^{-4}
0.6	1.1×10^{-2}	1.0×10^{-4}	1.1×10^{-2}
0.7	0.49	7.0×10^{-4}	0.49

The generation–recombination term is dominant below 0.3 V, and minority carrier injection is the major term above 0.4 V. The transition is quite pronounced on a plot of $\log_{10} I$ against V_A, so the approximation:

$$I = I_0' \exp \frac{qV_A}{nkT} \; \text{A} \tag{7.27}$$

is only valid over a restricted range. The ideality factor n and the pre-exponential current I_0' can be found from the logarithmic plot. Equation (7.25) becomes:

$$\log_{10} I = \log_{10} I_0' + \frac{qV_A}{nkT} \log_{10} e$$

Hence the slope is:

$$\frac{\mathrm{d}}{\mathrm{d}V_A} (\log I) = \frac{q \log_{10} e}{nkT}$$

which gives the values of n for the three regions as:

$n = 2$ (low forward bias)
$n = 1.3$ (intermediate bias)
$n = 1$ (high forward bias)

The intercept at $V_A = 0$ gives the value of $\log I_0'$ as -10.15, so that $I_0' = 7.1 \times 10^{-11}$ A. It should be noted that I_0' has no physical significance, unlike I_0 and I_{GR0}. It is simply a parameter which gives the best fit between an empirical law and one based on physical principles.

7.4 Diode types

7.4.1 Rectifiers

Electrical power generation and transmission predominantly uses alternating current but a good deal of electrical equipment requires direct current. The first use of the *pn* junction is therefore as a power diode, normally in a full wave rectification circuit. This makes use of both half-cycles of the ac waveform and is easier to smooth to a steady dc level. The diode array can be made up from discrete diodes or can be obtained in a single package. The main design parameters are:

1. Forward current I_F, specified either as an average or a peak value, typically in the range 1–40 A.

2. Forward voltage V_F, at the rated forward current. Typically 0.7–1.5 V.

3. Maximum repetitive reverse voltage V_{RRM} which can be withstood without damage. High current devices are available with V_{RRM} up to about 1 kV. Extremely high tension (EHT) diodes exist with V_{RRM} of 25 kV, but are only rated at 3 mA.

4.   Reverse recovery time t_{rr}. This is unimportant for rectifying 50–60 Hz ac, where the diode capacitive impedance is very high, and is rarely quoted. Fast recovery rectifiers designed for use in switched mode power supplies must be able to respond rapidly, and have t_{rr} in the range 30–500 ns.

Rectification diodes dissipate a good deal of power. A dc calculation gives:

$$P = I_F V_F \text{ W} \tag{7.28}$$

which could be up to 60 W. The physical construction of these diodes is dominated by the need to minimize the thermal resistance R_{th} (in $°C W^{-1}$) between the junction and a distant, constant temperature ambient. High current diodes are stud mounted, so that one terminal takes the form of a threaded rod which can be clamped to a metal heatsink. The junction temperature is $\Delta T \, °C$ above ambient temperature, given by:

$$\Delta T = R_{th} P \, °C \tag{7.29}$$

R_{th} can be the sum of thermal resistances, for example between the junction and the case, between the case and the heatsink, and between heatsink and air. R_{th} is typically $20 \, °C W^{-1}$ for an axial wire leaded component, and $1 \, °C W^{-1}$ for a stud-mounted diode. The junction temperature itself must not exceed about 160 °C for silicon devices. Since equation (7.29) predicts the temperature rise of the junction above ambient, power diodes must be derated to lower maximum working currents at ambient temperatures above 50 °C or so. Designers should remember that ambient here means the temperature within the equipment box, not the air-conditioning setting.

EXAMPLE

■   A stud-mounted rectifier has a semi-ideal characteristic with a turn on voltage of 0.9 V. The junction to case thermal resistance is $1.5 \, °C W^{-1}$, and the heatsink is rated at $4 \, °C W^{-1}$. Plot a graph of the maximum steady forward current against ambient temperature, if the maximum junction temperature is 200 °C and the device is rated at 20 W. Assume perfect thermal contact between the diode and the heatsink.

☐   The diode power dissipation in forward bias is:

$$P = VI \text{ W}$$

$$= 0.9I \text{ W}$$

The total thermal resistance R_{th} is the sum of the series elements:

$$R_{th} = 4 + 1.5 = 5.5 \, °C W^{-1}$$

Hence the temperature *rise* of the junction above ambient is $T_j - T_a$:

$$T_j - T_a = R_{th}P = 4.95I \text{ °C}$$

If the maximum junction temperature T_j is 200 °C:

$$I_{max} = \frac{1}{4.95}(200 - T_a)$$

This sloping line is called the derating line, because it predicts that the maximum current is progressively reduced as the ambient temperature T_a increases. The other constraint is the maximum power of the device, 20 W, which sets an absolute maximum for I:

$$I_{max} \le P_{max}/V = 22 \text{ A} \quad (\text{all } T_a)$$

The device must be operated within the shaded region of Figure 7.14. The highest ambient temperature for operation at full power rating is 90 °C.

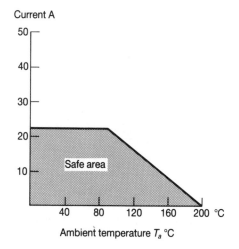

Figure 7.14 Derating of power diode at high temperature

7.4.2 Signal diodes

A diode designed to modify the flow of electronic information in a low power circuit is called a signal diode. The same *pn* junction theory applies to both these and rectifiers but the doping levels, size and packaging are chosen to produce a device with suitable electrical characteristics at the lowest cost. The same design parameters apply, except that power dissipation is rarely a problem in signal diodes. Typical values are:

$$I_F = 10\text{--}500 \text{ mA}$$
$$V_F = 0.4\text{--}1.9 \text{ V}$$
$$V_{RRM} = 10\text{--}100 \text{ V}$$
$$t_{rr} = 1 \text{ ns}$$

While silicon dominates the rectifier market, some signal diodes are made from germanium. Applications include high-speed switching, detectors and demodulators. The earliest radio sets used another kind of diode (the point contact diode) to rectify an amplitude-modulated (AM) radio frequency signal and recover the modulation envelope. Although the diode type has altered, the same principle is used to detect AM signals today.

7.4.3 Zener and voltage reference diodes

Reverse bias breakdown mechanisms were discussed in section 6.7. Zener diodes are designed to work as reference voltage elements in non-destructive reverse breakdown. As seen in Figure 7.15, the voltage across the diode stays nearly constant if the current is varied over a wide range. Design parameters include:

1. *Zener voltage, V_z*: this is the nominally constant reverse breakdown voltage. Devices are available with values from 3–75 V.

2. *Zener current, I_z*: the reverse current used to obtain V_z, typically 5–20 mA. When used as a voltage reference a current of this order must be fed through the diode in the *reverse* bias direction.

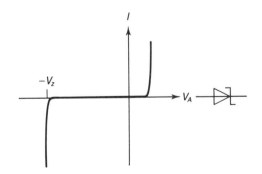

Figure 7.15 Zener diode characteristic and circuit symbol

3.   *Power rating*: a power of V_{zx} (reverse current) W is dissipated at the junction. Axial lead components can handle between 0.5 and 5 W, while stud-mounted zeners are available for higher power applications.

4.   *Slope resistance, r_e Ω*: the maximum slope resistance in reverse breakdown is usually quoted, in the range 1–100 Ω. This governs the stability of the zener voltage in response to a change in current I:

$$\Delta V_z = r_e \Delta I \text{ V} \tag{7.30}$$

Thus a 4.7 V zener with a slope resistance of 20 Ω will rise to 4.9 V if the rated zener current of 10 mA is doubled.

5.   *Temperature coefficient, $(\text{mV} \, °\text{C}^{-1})$*: the zener voltage varies with temperature quite markedly for high voltage components. The temperature coefficient varies with nominal V_z being negative for very low values of V_z, and positive at high V_z. The very low drift of V_z with temperature at 4.7 V accounts for the popularity of this value.

Other forms of reference voltage can be derived from semiconductor devices. The simplest is the drop across a forward biased diode, which is only available over a narrow band of values about 0.7 V. It is not difficult to show, using equation (6.25) for I_0 and equation (6.27) for the diode law, that at constant current V_A varies with temperature at:

$$\left. \frac{\Delta V_A}{\Delta T} \right|_{\text{constant } I} = \frac{q V_A - 3kT - E_g}{qT} \text{ V} \, °\text{C}^{-1} \tag{7.31}$$

This is negative because $q V_A$ is always less than the band gap. A silicon diode at 300 K has a forward-biased voltage/temperature coefficient of about $-2 \, \text{mV} \, °\text{C}^{-1}$ at $V_A = 0.7$ volts, provided the current remains constant.

A more sophisticated reference voltage is given by the Widlar band gap reference circuit. This relies on the fact that the *difference* in V_A for two identical diodes carrying different currents can possess a positive temperature coefficient. If diode $D1$ passes current I_1 with voltage V_{A1}, and $D2$ passes I_2 with V_{A2} ($I_1 > I_2$), then:

$$I_1 \simeq I_0 \exp \frac{q V_{A1}}{kT} \tag{7.32}$$

$$I_2 \simeq I_0 \exp \frac{q V_{A2}}{kT} \tag{7.33}$$

Taking the ratio:

$$\frac{I_1}{I_2} = \exp \frac{q}{kT} (V_{A1} - V_{A2}) \tag{7.34}$$

or:

$$(V_{A1} - V_{A2}) = \frac{kT}{q} \ln \frac{(I_1)}{I_2} \qquad (7.35)$$

If this difference is added in series with a proportion of the voltage across a third forward biased diode, the total voltage can have a temperature coefficient as low as $20\ \mu V\,°C^{-1}$. A full analysis shows that the output voltage with the smallest temperature coefficient is equal to the bandgap of the semiconductor used. Circuit techniques exist to derive other stable voltages from the bandgap reference.

7.4.4 Varactor diodes

The variation in small signal capacitance with voltage displayed by a reverse biased diode can be useful in automatic tuning circuits. This is far simpler and cheaper than the alternative motor-driven moving vane capacitor. Diodes designed for this use are called varactors (variable reactors) or tuning diodes. Although the capacitance is small varactors can be connected in parallel to provide higher values. Figures of importance include:

1. Capacitance variation with voltage: typically 2–20 pF for reverse voltages between −20 and −0.5 V.
2. Reverse breakdown voltage: −30 V.
3. Reverse leakage current: 0.5 μA.

Summary

Main symbols introduced

Symbol	Unit	Description
C_d	$F\,m^{-2}$	small-signal diffusion capacitance
C_j	$F\,m^{-2}$	small-signal junction capacitance
n	—	diode ideality factor
R_{th}	$°C\,W^{-1}$	thermal resistance
r_e	Ω	small-signal resistance
t_s	s	storage time in diode switching
t_{rr}	s	reverse recovery time
V_{RRM}	V	maximum repetitive reverse voltage without diode breakdown
V_z	V	zener breakdown voltage

Main equations

$$r_e = \frac{kT}{qI} \quad \text{(forward bias)} \tag{7.4}$$

$$C_j = -dQ_j/dV_A \tag{7.7}$$

$$C_d = dQ_d/dV_A \tag{7.12}$$

Exercises

1. Calculate the dc and small-signal ac resistances of a *pn* diode at 27 °C with a forward current of 50 mA and the following properties: $n_i = 10^{16}$ m^{-3}, $N_A = N_D = 10^{21}$ m^{-3}, $W_p = W_n = 0.02$ mm, $\mu_e = 0.1$, $\mu_n = 0.05$ m^2 V^{-1} s^{-1}, cross-sectional area 1.5 mm^2.

2. A varactor diode of 2 mm^2 cross-sectional area is to be constructed in silicon as a wide, single-sided, abrupt p^+n junction. Suggest doping levels for each side if the small-signal capacitance is to be at least 15 pF at 1 V reverse bias. Name two other parameters influenced by the doping on the lightly doped side.

3. A linear junction is formed at $x = 0$ by constant acceptor doping N_A and a linear variation in N_D: $N_D(x) = ax + N_A$. Find the effective doping levels each side of the junction. Solve Poisson's equation through the junction region using the depletion approximation and the fact that $d_n = d_p$ (by symmetry). Hence show that the junction capacitance is proportional to $V_t^{-1/3}$ in this case (a is a positive constant).

4. Find the ratio of the small-signal resistive and capacitative impedances of the diode in (1) at 1 MHz, assuming $L_e = L_n = 0.1$ mm and $V_A = 0.4$ V. At what frequency are the impedances equal?

5. A wide silicon diode has the following properties: $N_A = 10^{25}$ m^{-3}, $N_A = 10^{20}$ m^{-3}, area 2 mm^2, $I_0 = 3$ nA, $\mu_e = 0.14$ m^2 V^{-1} s^{-1}, $\tau_e = 5$ μs, at 300 K. Find the junction capacitance at 3 V reverse bias and the diffusion capacitance at 0.4 V forward bias.

6. The storage time t_s is measured by applying a square wave of $\pm V$ volts to a diode in series with a resistor. The steady forward current I_f during the positive half cycle and the peak reverse current I_r during the storage phase are also measured. The results are:

V	volts	1	2	3	4	5
I_f	mA	1.4	5.9	10.5	15.0	19.5
I_r	mA	7.7	12.3	16.8	21.4	25.9
t_s	μs	3.2	7.9	9.7	10.6	11.2

Deduce a value for the minority carrier lifetime.

7.   Determine values for the pre-exponential current I_0' and the ideality factor n for a silicon diode at 300 K in which $A = 3$ mm^2, $N_A = N_D = 10^{23}$ m^{-3}, $\mu_e = 0.14$ m^2 V^{-1} s^{-1}, $\mu_h = 0.05$ m^2 V^{-1} s^{-1}, $\tau_e = \tau_h = 15$ μs. Over what range of applied voltage is the empirical model accurate to within 25%?

8.   A piece-wise linear power rectifier rated at 10 W has a turn on voltage of 0.6 V and a constant slope resistance of 0.5 Ω. The junction-case thermal resistance is 2 °C W^{-1}, its heatsink is rated 5 °C W^{-1}, and the maximum junction temperature is 150 °C. Plot a graph of maximum dc current against ambient temperature. Could the diode be run at full power at 100 °C ambient temperature with a different heatsink?

9.   A zener diode is rated $V_z = 5.1$ V at 20 °C ambient with $I_z = 40$ mA. The temperature coefficient of V_z is 2 mV °C^{-1}, the slope resistance is 20 Ω, and the thermal resistance between the junction and ambient is 40 °C W^{-1}. Calculate the derating in mW °C^{-1} necessary to maintain the zener voltage constant at higher ambient temperatures. Find also the upper and lower limits of I_z for V_z to stay within $\pm 5\%$ of the specified value, assuming $T_a = 20$ °C.

10.   Derive equation (7.31) for the temperature coefficient of the forward bias voltage at constant current. Evaluate it at 300 K and 77 K for germanium and gallium arsenide for $qV_A = E_g/2$.

8 | Optoelectronics

Review of fundamentals

We receive more information through our eyes than by any other sense. Optical signalling has advanced from the beacon fires which heralded an enemy's approach to being the fastest communication medium available. Information processing, however, is still best carried out electrically in semiconductor devices. This combination has led to the widespread use of optoelectronic devices and systems in which both light and voltage signals are transferred, and in this chapter we shall mainly be concerned with the transducers which can convert between the two media.

Light is electromagnetic radiation with wavelengths from 390–780 nm, which is visible to the human eye. Many optoelectronic devices work in the near infrared, in the range 780–3,000 nm. What, then, is the origin of light? When an electron makes a rapid transition between energy levels the fact is announced to the outside world by the emission of a photon, a packet of electromagnetic radiation. Examples of this are found in electric plasmas, such as sodium vapour or neon lamps, in which free electrons give energy to attached electrons in collisions. The bound electron is excited to a higher permitted level and can fall back to its original state with the emission of a photon. We recall from Chapter 3 that the energy of a photon E_p in terms of the wavelength λ or frequency v of the radiation is:

$$E_p = hv = hc/\lambda \tag{8.1}$$

where c is the speed of light and h is Planck's constant. Transitions between permitted bands of levels in a semiconductor in which electron–hole pairs are created and recombine can also produce photons, provided a direct transition can be made. This was discussed in more detail in section 5.1.

Radiation with a more continuous character is emitted by all bodies by virtue of their temperature. One of the earliest triumphs of quantum theory was to provide a theoretical explanation for the distribution of thermal radiation with wavelength from a perfectly absorbing surface – the black-body spectrum shown in Figure 8.1. This set of curves matches the common observation that hotter objects are both brighter and take a colour further from the red end of the spectrum. The formula derived by Planck in 1900, first empirically to fit measured data, and then theoretically, is:

$$M = \frac{2\pi hc^2}{\lambda^5} \frac{1}{\exp\left(hc/\lambda kT\right) - 1} \tag{8.2}$$

M is the spectral flux density, or spectral exitance, in watts per square metre per unit wavelength interval. The area under each curve in Figure 8.1 is the total power density emitted in $W\,m^{-2}$, also known as the exitance. This varies with the fourth power of absolute temperature, a relation known as the Stephan–Boltzmann law:

$$P/A = \sigma_{SB} T^4 \; W\,m^{-2} \tag{8.3}$$

The Stephan–Boltzmann constant σ_{SB} is $5.67 \times 10^{-8}\,W\,m^{-2}\,K^{-4}$. This applies to perfect black bodies. Real surfaces reflect a fraction $(1 - e)$ of the incoming radiation and the righthand side of the Stephan–Boltzmann law must be multiplied by the emissivity e. The law fails if the emissivity varies with wavelength, as it does in the case of silicon. (This poses a particular problem for remotely sensing the temperature of a piece of silicon during device fabrication.)

A quantity of interest to both emitting and absorbing devices is the absorption per unit length α at a given wavelength. If α is very high nearly all the incident radiation will be absorbed close to the surface, and most of the electron–hole pairs produced will be lost by rapid surface recombination. Should α be very low there will be little light lost in transmission through the device, which is good for an emitting device but bad for a detector. The variation of intensity through a medium is derived by considering the change in light flux

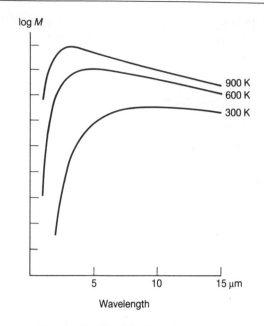

Figure 8.1 The black-body spectrum

(F photons $\mathrm{m}^{-2}\,\mathrm{s}^{-1}$) in a small distance $\mathrm{d}x$ within the material (Figure 8.2). The fractional change in F is $\alpha \mathrm{d}x$:

$$\frac{\mathrm{d}F}{F} = -\alpha \mathrm{d}x \tag{8.4}$$

The minus sign is because the photon flux is falling. Integrating with limits $F = F_0$ at $x = 0$ and a value F at position x:

$$F = F_0 e^{-\alpha x} \tag{8.5}$$

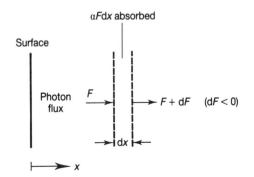

Figure 8.2 Light absorption

The variation of α with wavelength for some semiconductor materials is given in Figure 8.3. Direct gap materials exhibit a sharp transition in absorption at the band gap energy, while indirect gap materials are much less abrupt, reflecting the low probability of a direct transition at the minimum energy.

It is easy to be confused by the variety of units for light measurement, so we distinguish two kinds:

1. Absolute units, for light of a specified wavelength, such as the total power (W), the photon flux (photon s^{-1}), and the power density (W m^{-2}) and photon flux density (photons $m^{-2} s^{-1}$).

2. Units weighted to the eye response, Figure 8.4. This has a peak in the green at 555 nm. One watt of power at this wavelength produces a total *luminous flux* of 680 *lumens*. The same power if emitted at other wavelengths produces a smaller response from the eye, and thus possesses a smaller luminous flux. The electrical and eye-response efficiencies of lighting sources are often combined in an overall luminous efficiency, measured in lumens per watt. For example a fluorescent tube has a typical luminous efficiency of 40 lumens/watt. When considering the amount of light falling on a surface, the illumination is measured in lumens per square metre, or *lux*.

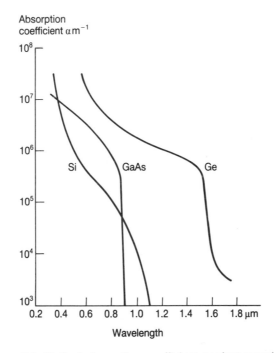

Figure 8.3 Optical absorption coefficient against wavelength

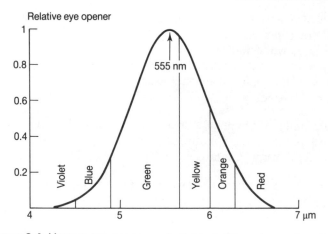

Figure 8.4 Human eye response to light of different wavelengths

EXAMPLE

■ The active layer of a silicon photodector begins 1 micron below the surface and has a thickness of 10 microns. Determine the fraction of light intercepted if the absorption coefficient is 10^5 m^{-1}.

□ One micron below the surface the light flux has fallen to:

$$\exp-(10^5 \times 10^{-6}) = 0.905$$

of its value at the surface. At the bottom of the active layer this has fallen to:

$$\exp-(10^5 \times 11 \times 10^{-6}) = 0.333$$

of the surface value. The fraction absorbed within the active layer is $0.905 - 0.333 = 0.572$. Nearly 10% is absorbed before reaching the active layer and 33% is transmitted through the device without being absorbed.

8.2 Photoconductors

One way of detecting the level of light is to use the change of conductivity which occurs when light frees extra mobile carriers. We saw in Chapter 5 that an excess generation $\hat{G}$ m^{-3}s^{-1} of electron–hole pairs produces excess carrier densities $\hat{n}$ and $\hat{p}$ given by:

$$\hat{n} = \hat{G}\tau_e \text{ m}^{-3} \quad (p\text{-type})$$

$$\hat{p} = \hat{G}\tau_h \text{ m}^{-3} \quad (n\text{-type})$$

(8.6)

where τ_e and τ_h are the electron and hole lifetimes respectively. The change in conductivity $\hat{\sigma}$ is therefore:

$$\hat{\sigma} = q(\mu_e \hat{n} + \mu_h \hat{p}) = q\hat{G}(\mu_e \tau_e + \mu_h \tau_h)\ \Omega^{-1}\,m^{-1} \tag{8.7}$$

Table 8.1 lists the properties of photoconductors in common use at room temperature. Other compounds and alloys are suitable for detecting 8–14 μm wavelength radiation, notably mercury–cadmium telluride (Hg/CdTe) and lead –tin telluride (Pb/SnTe). Devices which use these materials must normally be cooled to 77 K by liquid nitrogen to reduce the intrinsic conductivity. This makes smaller changes in conductivity detectable. These wavelengths are used for infrared night sights because black-body emission at 300 K peaks around 10 μm and there is little atmospheric absorption at this wavelength. A more exotic class of extrinsic photodetectors operate at even lower temperatures, below 20 K. These rely on thc radiation freeing charges from *dopants* into the conduction band, rather than a band to band transition. The very low temperatures are necessary to keep a significant fraction of dopants unionized in the absence of radiation.

Consider now the simple photoconductor and circuit in Figure 8.5. The illuminated surface area is wl m^2, on which falls a photon flux F m^{-2}s^{-1}, neglecting reflection. Let us assume that the depth d is sufficient to absorb virtually all the photon flux. The excess generation rate averaged over the depth is therefore:

$$\hat{G} = \eta F/d\ m^{-3}\,s^{-1} \tag{8.8}$$

where η is the quantum efficiency, which is the number of electron–hole pairs created per photon (usually one). With a voltage V applied across a length l of photoconductor, the photocurrent I_{ph} which flows is therefore:

$$I_{ph} = wd\hat{\sigma}V/l = Vwq\eta F(\mu_e \tau_e + \mu_h \tau_h)/l\ A \tag{8.9}$$

This can be compared with the charges generated per second within the

Table 8.1 Photoconductor properties at 300 K

Material	Band gap (eV)	Maximum wavelength absorbed (μm)	Electron mobility (μ_e m^2 V^{-1} s^{-1})
Ge	0.67	1.8	0.39
Si	1.12	1.1	0.14
GaAs	1.42	0.87	0.85
GaP	2.26	0.55	0.03
CdS	2.42	0.51	0.03
ZnS	3.68	0.34	0.02

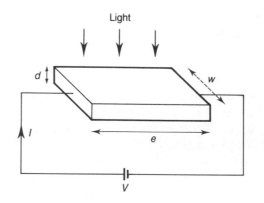

Figure 8.5 Photoconductor

photoconductor, which can also be expressed as a current I'_{ph} by multiplying by the charge per particle, q:

$$I'_{ph} = q\hat{G}wdl \ \text{A}$$
$$= q\eta Fwl \ \text{A} \tag{8.10}$$

The ratio of I_{ph} and I'_{ph} is called the *gain* g_{ph} of the photoconductor:

$$g_{ph} = I_{ph}/I'_{ph}$$
$$= V(\mu_e\tau_e + \mu_h\tau_h)/l^2 \tag{8.11}$$

A high photocurrent gain requires high mobility, long carrier lifetime, and a short distance l between the contacts. The gain can also be expressed in terms of the transit time t_t across the device:

$$t_t = l^2/\mu_e V \tag{8.12}$$

so if the hole current can be neglected ($\mu_h \ll \mu_e$),

$$g_{ph} = \frac{\tau_e}{t_t} \tag{8.13}$$

A high gain may be achieved at the cost of a slow response time for the device. Its ability to detect repetitive pulses is restricted to those of frequency $f \ll 1/\tau_e$. Another parameter of importance is the *dark current* I_{pho} which flows in the absence of light, and should be as low as possible. Photoconductors, also known as *light dependent resistors*, are still the best detectors in the infrared above a few microns wavelength, in spite of competition from junction-type detectors. Typical figures are:

gain g_{ph} 1–10^6

response time 0.1–10^{-8} s (mostly long)

EXAMPLE

■   Find the gain of a photoconductor 0.1 mm between contacts with an applied field of $4.10^5 \, \mathrm{V\,m^{-1}}$, electron mobility $0.14 \, \mathrm{m^2\,V^{-1}\,s^{-1}}$ electron lifetime 10^{-5} seconds.

☐   The transit drift velocity is the product of the mobility and field:

$$v_d = 0.14 \times 4 \times 10^5 = 0.56 \times 10^5 \, \mathrm{m\,s^{-1}}$$

The transit time is therefore $l/v_d = 1.79 \times 10^{-9}$ s. The gain, according to equation (8.13) is $10^{-5}/1.79 \times 10^{-9} = 5{,}600$.

8.3 PIN diode detectors

We found in section 7.3 that if an electron–hole pair was created in the depletion region of a reverse biased *pn* junction the charges were quickly separated by the field at the junction and flowed around the external bias circuit, contributing to the reverse leakage current. If all we are trying to do is build a perfect rectifier this is merely a nuisance, but it can be turned to advantage as a means of detecting light. There are two requirements:

1.   The light must be of sufficiently short wavelength to promote electrons to the conduction band.
2.   Absorption must take place in the depletion region.

The second condition is difficult to meet at first sight because several tens of microns are often needed to absorb light; but depletion region widths are usually no bigger than a micron. The answer is to tailor the doping to stretch out the depletion region. One way would be to construct a $p^+ n^{--}$ diode, so that the depletion region would extend a long way into the lightly doped *n*-side (Figure 8.6a). This design has the disadvantage of a high series resistance within the device. If three layers of doping are used: $p^+ i n^+$, where *i* means intrinsic, the depletion region can extend right the way across the central region (Figure 8.6b). In practice a very lightly doped central region is used. The reverse bias current of this PIN diode structure can be used to measure light, if it is encapsulated with a window.

A popular construction is shown in Figure 8.7. The top contact is either a transparent conductor (indium/tin oxide is suitable for visible light), or else consists of fingers which leave the maximum amount of diode surface visible without increasing the contact resistance unduly. The *p*-type layer is made thin because electron–hole pairs created more than a diffusion length from the

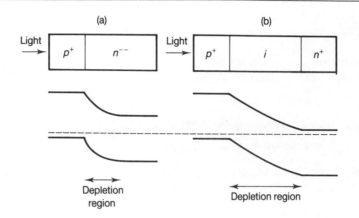

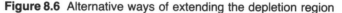

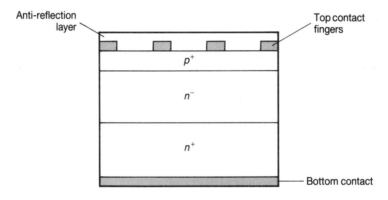

Figure 8.6 Alternative ways of extending the depletion region

Figure 8.7 PIN diode detector construction

depletion region will recombine locally without adding to the current. The *p*-layer must therefore be heavily doped to avoid punchthrough breakdown.

The main design parameters for a diode photodetector are:

1. *Wavelength sensitivity*: usually given as a peak response wavelength or a graph.

2. *Response*: this gives the change in reverse current for a given increase in total power (or power per unit area) of light of specified wavelength falling on the device, typically $10 \, \text{A} \, \text{W}^{-1} \, \text{m}^{-2}$.

3. *Dark current* (A): the reverse saturation current which flows at a specified reverse bias without incident light, typically 1–500 nA. This determines the

minimum light level which can be sensed without using a differencing technique.

4. *Capacitance* (pF): at a specified reverse bias voltage, typically 4–400 pF, this together with the resistance of the device and the external circuit, determines the risetime of the photocurrent.

5. *Reverse breakdown voltage* (V), typically 50–100 volts: some devices are operated close to breakdown, so that pulses of photocurrent are multiplied by avalanche. This must be avoided if a linear response is required.

8.4 PIN solar cells

The dream of harnessing the sun's power has captured mankind's imagination for centuries, and with good reason: up to 800 W of free power falls on every square metre at noon. So far the cost of solar cells has limited their application to space vehicles and remote terrestrial locations. More efficient means of energy storage will also be required to cope with daily and seasonal variations in both sunshine and demand for power.

Much the same design considerations apply to photodetector diodes and solar cells. Both are optimized for electron–hole pair creation in the depletion region, with a low series resistance. The photocurrent I_{ph} reinforces the reverse saturation current I_0, and flows in both forward and reverse bias. The I/V characteristic is thus a small change to the diode law:

$$I = I_0\left(\exp\left(\frac{qV_A}{kT}\right) - 1\right) - I_{ph} \text{ A} \tag{8.14}$$

which is plotted in Figure 8.8. The normal diode characteristic is shifted down

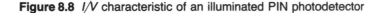

Figure 8.8 *I/V* characteristic of an illuminated PIN photodetector

by a constant amount, if the variation in depletion region width with bias in the PIN structure can be neglected. While the diode law passes through the origin, the photodiode or solar cell is characterized by two quantities, for a given illumination:

1.   The open circuit voltage V_{oc}: if the diode is disconnected, a forward bias voltage appears across its terminals. In order to make the total current zero, a forward current of I_{ph} must flow internally to counter the photocurrent of $-I_{ph}$. The separation of charge which occurs when electron–hole pairs are formed in the depletion region tends to lower the junction barrier, permitting a balancing diffusion current to flow. The maximum value for V_{oc} is the built in potential V_B, which is generally smaller than the band gap potential.

2.   The short circuit current I_{sc}. If $V_A = 0$ is inserted in equation (8.14), the current is $-(I_{ph} + I_0) \simeq -I_{ph}$, also known as the short circuit current. The diode in zero or reverse bias acts as a current source to a low impedance load, as the photo-generated excess carriers pass round the external circuit to counter the charge imbalance caused by their separation. This is the mode of operation used for photodetectors.

Now suppose that an illuminated solar cell is connected across a load resistance R, as in Figure (8.9a). How does this act as a power source? We will consider the problem in two ways. First look at the I/V characteristic. In the first quadrant both V_A and I are positive, and the power dissipated in the diode is IV_A, which is positive. Similarly in the third quadrant, both V_A and I are negative, making the product positive once more. The normal diode is limited to these two regions, but the solar cell characteristic passes through the fourth quadrant. Here V_A is positive while I is negative, so the power dissipated in the diode is negative, which means that power flows *out* of the diode into the external circuit.

An alternative viewpoint is to consider the motion of the charges (Figure 8.9b and c). The depletion region field draws electrons to the n-side and holes to the p-side. The charges are drawn back together around the external circuit, and thus flow in the same sense as charges driven by a battery, with the p-side corresponding to the positive terminal. Energy is therefore transferred from the light to the load resistor and the diode acts as a light-driven battery.

To analyse such a circuit it is customary to draw a load-line on the I/V characteristic. This is electronic engineering jargon for the solution of two simultaneous equations by means of a graph. The unknown quantities are the current I and the diode voltage V_A. One equation linking these two is equation (8.14) the photodiode law. The current in this equation is positive for positive charges moving through the diode from the p-side to the n-side. A comparison with the battery equivalent circuit shows that this drives current in the negative

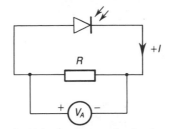

(a) Solar cell with load resistance, showing the positive
senses of voltage and current

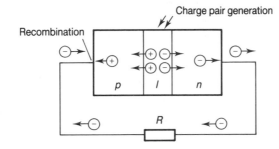

(b) Charge motion around the circuit

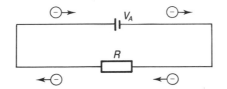

(c) Equivalent current driven by a battery. V_A is positive
and I is negative

Figure 8.9 Solar cell

direction, from p- to n-side around the external circuit. Ohm's law for the
external circuit is therefore:

$$V_A = -IR \qquad\qquad (8.15)$$

This constitutes the second condition which must be satisfied. If the straight-line
relation, equation (8.15), is plotted on the diode I/V characteristic graph, it is
then called a load-line (Figure 8.10). Both the diode law and the circuit are
satisfied at the point of intersection. (This is a general method and can be
applied to many other circuits and devices.)

Power transferred to the load resistor is the product of V_A and I, which is
equivalent to the area of the shaded rectangle. If R is either too big or too small

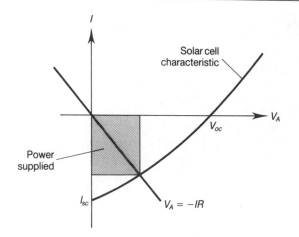

Figure 8.10 Load-line method for finding V_A and I for a loaded solar cell

the power transfer will be very small. Solar cells can be connected in series or parallel to increase the voltage and current to useful values for a given load.

Figures of importance for individual solar cells include:

1. Surface area (m^2).
2. Efficiency of energy conversion. Values in the range 10–15% are typical, although theoretical maximum efficiencies are higher.
3. Open circuit voltage V_{oc} (0.5–0.6 V for silicon).
4. Short circuit current I_{sc}.
5. Temperature coefficient of efficiency. The efficiency falls almost linearly with an increase in temperature, up to about 200 °C for silicon and 300 °C for GaAs.
6. Cost per watt. The price should figure in *all* assessments of the relative merits of devices (as should the projected delivery date), but we highlight it here because of the key role of economics in the growth of the solar cell market. While it is cheaper to generate electrical power by burning fossil or nuclear fuels, solar cells are limited to a narrow market. Should the price advantage change, solar cell manufacturers will be sitting on gold-mines.

Some of the measures being researched to increase efficiency and reduce cell costs include:

1. Anti-reflection coatings and textured crystal surfaces, to increase the amount of light entering the device.
2. Heterojunction and multi-junction designs, using *pn* junctions made from dissimilar semiconductors. Heterojunctions have potentially lower leakage

currents, while stacking semiconductors of different band gaps can make better use of the energy in the solar spectrum.

3. Thin film solar cells can be deposited relatively cheaply on low cost substrates. This avoids using expensive high grade semiconductor material simply for mechanical support. Cadmium sulphide and amorphous silicon cells are examples of this design.

4. Mirror concentrators. Solar cells increase in efficiency with the intensity of radiation, provided their temperature is controlled. There may be an advantage in using a sun-tracking optical system to focus light on to a small, high quality, cell.

8.5 Light emission

We have noted already that light is produced when an electron makes a rapid transition from a higher to a lower energy state. Radiation due to thermal energy is emitted by all bodies, and was considered in section 8.1. Many materials are luminescent – capable of emitting useful quantities of light – and their properties are reviewed briefly in this section. Types of luminescence are distinguished by the source of primary excitation:

1. *Photoluminescence*: high energy photons are absorbed and lower energy photons emitted.

2. *Electroluminescence*: the passage of current through a material or junction causing light emission.

3. *Cathodoluminescence*: the impact of an electron beam (a cathode ray) excites emission.

4. *Radioluminescence*: emission caused by other fast particles or radiation.

Apart from semiconductors, by far the most important group of luminescent materials are the compounds of phosphorus known as phosphors. The very name of the element means light bearing and was also the name given to the planet Venus in antiquity. Luminescence in phosphors is a three-stage process (Figure 8.11). First, an electron is excited to a higher energy level by one of the above mechanisms. Second, there is a rapid transition to an intermediate level, usually without visible photon emission. The position of this level determines the energy of the photon emitted when the electron finally returns to its ground state. The decay may be very rapid (less than 0.1 microseconds) in which case the whole phenomenon is called *fluorescence*, or it may extend to seconds, minutes or even hours in which case it is termed *phosphorescence*.

Cathodoluminescent phosphors are used in oscilloscopes and television tubes. The addition of different impurities gives rise to the red, green or blue light from adjacent phosphor dots, which create colour television pictures. The rate of decay must be balanced between too short, which will give a flickering

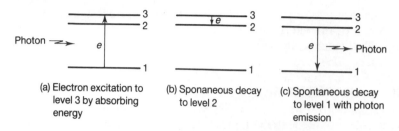

Figure 8.11 Light emission in phosphors

image, and too long, which will blur a moving picture. Photoluminescent phosphors are used in fluorescent tube lights, to convert the strong ultraviolet light from the central mercury vapour/argon glow discharge into visible light. The art in this case is to achieve the desired colour as well as a high conversion efficiency. Such lights have some of the highest luminous efficiencies among types in common use.

8.6 Light-emitting diodes (LEDs)

Electroluminescence is the most important light emission mechanism in semiconductor devices. In a direct band gap material, recombination of electron–hole pairs is accompanied by photon emission, so all we need for light production is a surplus population of minority carriers. Once again we turn to the *pn* junction diode: in forward bias the minority carrier population each side of the junction is strongly increased. The brightness of an LED is therefore controlled by the forward current. Insufficient light is produced for illumination but LEDs are widely used as indicator lights because they can be operated at low voltages and have a longer life than filament lamps.

Colour, or wavelength, is the most obvious property of an LED, and is determined according to equation (3.4) by the band gap of the material used. The ternary (three-component) compound semiconductor gallium arsenide-phosphide is of particular interest. Arsenic and phosphorus both possess five valence electrons and can form a stable compound semiconductor with trivalent gallium, if tri- and pentavalent atoms occupy alternate crystal sites. This means that a whole range of semiconductors of composition $GaAs_{1-x}P_x$ can be formed, where x is $\leqslant 1$, from GaAs ($x = 0$) to GaP ($x = 1$), with band gaps between 1.43 eV and 2.26 eV (Figure 8.12). GaAs itself emits in the infrared, which is only useful for optical couplers and other emitter–detector applications such as optical shaft encoders (see below). Sadly, the gap becomes predominantly indirect for $x > 0.45$, but values just under this yield devices which emit red visible light. Useful emission in the yellow–green part of the spectrum has been obtained by

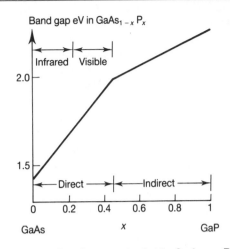

Figure 8.12 Band gap potential in GaAs$_{1-x}P_x$

doping GaAs$_x$P$_{1-x}$ $(x > 0.45)$ with nitrogen. This acts as a stepping stone for recombination, with the unusual property of allowing radiative recombination between a trapped electron and a hole. The photon energy corresponds to the trap energy above the valence band rather than the full band gap energy, and both depend on the alloy composition. Blue LEDs were elusive for many years, but have now been made using zinc sulphide (ZnS) or silicon carbide (SiC). They are currently more expensive than GaAsP types.

Figures of importance and typical values for LEDs are:

1. Peak emission wavelength: 635 nm (red), 585 nm (yellow), 565 nm (green).
2. Forward operating current I_F: 5–40 mA, typically 10 mA. The more current, the more light. Nitrogen-doped yellow and green types sometimes need a little more current than red LEDs.
3. Forward bias voltage at I_F: typically 2 V.
4. Intensity: 2–5 mcd (millicandelas). This must be multiplied by the response of the eye at different wavelengths in order to get the perceived brightness. Far more energy must be radiated at the ends of the visible spectrum (red and blue) to attain the same visual stimulus as a given energy of green light.
5. Viewing angle: 30–90°. The transparent cover encapsulating the LED acts as a lens. High brightness or low current versions concentrate the light into a smaller viewing angle.

LEDs can be created in bar or pinpoint arrays to display numbers or figures. These are tending to be displaced by low power liquid crystal displays in applications which are not operated in the dark.

8.7 Optical couplers

A light-emitting diode and a matched photodetector can be packaged together as in Figure 8.13 to make an optical coupler, also known as an optical isolator. The detector can be a PIN diode or a light-activated transistor (a phototransistor), in which the light controls a far larger detector current. There is no electrical connection between the input and output devices, not even a ground connection. Optical couplers can be used to send signals between low voltage circuitry and a subsystem which may be floating at a few kilovolts. They are also useful in isolating a signal source from a noisy load and in coupling logic families operating at different supply voltages. A silicon-doped GaAs emitter combined with a silicon photodiode detector has a recovery time as low as 2 ns and can be used for high speed transmission of digital data.

A key parameter is the current transfer ratio I_2/I_1. Typical values are 0.001–0.002 for an LED–PIN diode pair, and 10 for a phototransistor.

8.8 Semiconductor lasers

8.8.1 Laser fundamentals

The light from an LED is *incoherent*, which means that there is no ordered phase relationship in the electromagnetic wave emitted. It is also relatively broad band, the width of the half-power points being 10–20 nm. In this section we will discuss the conditions required to achieve the coherent, narrow band emission necessary for very high speed communication.

So far we have considered two electron-photon interactions:

1. *Absorption*, in which an electron is promoted to a higher energy with the loss of a photon:

$$ \underset{\text{electron}}{e} + \underset{\text{photon}}{h\nu} \longrightarrow \underset{\text{excited electron}}{e^*} \qquad (8.16) $$

2. *Spontaneous emission*, the reverse process:

$$ e^* \longrightarrow e + h\nu \qquad (8.17) $$

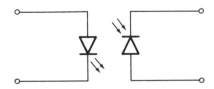

Figure 8.13 LED and photodiode linked only by light in an optical isolator

A third process also exists:

3.   *Stimulated emission*, in which a photon interacts with an excited electron and induces the transition to a lower state:

$$e^* + h\nu \longrightarrow e + h\nu + h\nu \qquad (8.18)$$

A remarkable property of stimulated emission is that the two photons are coherent, with oscillations exactly in phase with one another. Clearly this opens the way for photon amplification, *provided* stimulated emission can exceed spontaneous emission and absorption. To define this condition we will need to establish the relative probabilities of the events.

Consider the two-level system with energies E_1 and E_2 ($E_1 < E_2$) shown in Figure 8.14, with N_1 atoms at E_1 and N_2 atoms at E_2. The Boltzmann relation predicts that at equilibrium at temperature T:

$$N_2 = N_1 \exp \frac{-(E_2 - E_1)}{kT} \qquad (8.19)$$

Suppose now that this assembly of atoms is in equilibrium with black-body radiation at the same temperature. To get our final answer in the conventional form we need to convert from the spectral exitance (equation (8.2)) in watts m^{-2} per wavelength interval to the radiation density $R(\nu)$ in units of energy m^{-3} per frequency interval. This is accomplished by multiplying M by $4\lambda^2/c^2$, and expressing R in terms of the frequency $\nu = c/\lambda$:

$$R(\nu) = \frac{8\pi h\nu^3/c^3}{\exp(h\nu/kT) - 1} \qquad (8.20)$$

The probabilities of the three events are known as the Einstein coefficients (it's that man again!):

1. A_{21}: the probability per unit time of an excited atom decaying spontaneously.

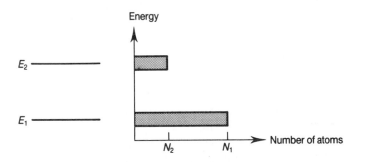

Figure 8.14 Populations in energy levels E_1 and E_2 at equilibrium

2. B_{12}: $B_{12}R$ is the probability per unit time of an atom in level one being excited to level 2 by absorbing a photon.
3. B_{21}: $B_{21}R$ is the probability per unit time of an excited atom decaying to state 1 by stimulated emission.

In equilibrium there is a balance between upward and downward transitions, which will also depend on the number of atoms in each state:

$$B_{12}N_1R \quad = \quad A_{21}N_2 \quad + \quad B_{21}N_2R \quad\quad\quad (8.21)$$

$$\text{absorption m}^{-3}\,\text{s}^{-1} = \quad \substack{\text{spontaneous} \\ \text{emission m}^{-3}\,\text{s}^{-1}} \quad + \quad \substack{\text{stimulated} \\ \text{emission m}^{-3}\,\text{s}^{-1}}$$

Using equation (8.19) this becomes:

$$R = \frac{A_{21}}{B_{12}\exp\dfrac{(E_2 - E_1)}{kT} - B_{21}} \quad\quad\quad (8.22)$$

This is identical to the black-body equation (8.20) if:

$$h\nu = E_2 - E_1$$

$$B_{12} = B_{21} \quad\quad\quad (8.23)$$

$$A_{21} = \frac{8\pi h\nu^3 B_{12}}{c^3}$$

The ratio of stimulated emission to spontaneous emission is:

$$\frac{B_{21}N_2R}{A_{21}N_2} = \frac{Rc^3}{8\pi h\nu^3} \quad\quad\quad (8.24)$$

which means that it is easier to observe stimulated emission at lower frequencies. This was one reason why microwave stimulated emission at $\nu \sim 10^9$ Hz was proved long before the stimulated emission of visible light at $\nu \sim 10^{15}$ Hz. A greater radiation density is required at higher frequencies if stimulated emission is to exceed spontaneous emission. A second requirement is that stimulated emission must be greater than absorption. The ratio of these terms is:

$$\frac{B_{21}N_2R}{B_{12}N_1R} = \frac{N_2}{N_1} \qu\quad\quad (8.25)$$

The Boltzmann relation (8.19) predicts that N_2 is always less than N_1 while the system is in equilibrium. If stimulated emission is to dominate the system must be removed from equilibrium so far that $N_2 > N_1$. This situation is known as *population inversion*. A sufficiently high radiation density and population inversion are the two preconditions for laser action. 'Laser' is an acronym for light amplification by stimulated emission of radiation.

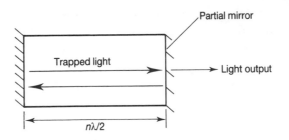

Figure 8.15 A simple laser cavity

Laser action is most easily achieved in a *resonant cavity* (Figure 8.15) filled with the emitting material. The ends are partial mirrors, reflecting the radiation back and forth to increase the radiation density. Light escaping through the partial mirrors is the useful output. The cavity size is chosen to be an integral number of half wavelengths of the characteristic light from the system. This encourages coherent emission from the whole assembly.

8.8.2 Junction lasers

It is possible to achieve population inversion in a forward biased *pn* junction between highly doped materials. The doping must be *degenerate*, which means that the Fermi level no longer resides within the band gap, and the assumptions previously used for calculating n and p fail. Figure 8.16a shows the zero bias band diagram, with the energy levels populated by electrons shaded. The depletion region is thin and the built-in potential is higher than normal,

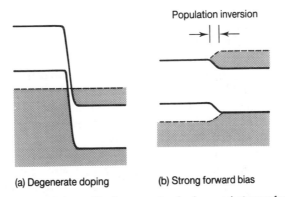

(a) Degenerate doping   (b) Strong forward bias

Figure 8.16 *pn* junction with degenerate doping and strong forward bias

exceeding the band gap potential. If sufficient forward bias can be applied (Figure 8.16b), conduction band electrons and valence band holes penetrate the depletion region in sufficient numbers to create a population inversion zone, with a large number of conduction band electrons coexisting with a large number of valence band holes. This meets one of the criteria for laser action.

The requirement of a high radiation density in a resonant cavity can be satisfied by a device construction like that in Figure 8.17. The mirror faces are made parallel by orienting the device with respect to the semiconductor crystal, so that proper cleaving can produce a laser cavity. Faces in the perpendicular direction are deliberately roughened to discourage other laser modes from building up. When the device is forward biased it behaves initially like an LED until, at a current density above threshold, coherent stimulated emission dominates. A steady state condition exists when the light lost by external emission and absorption balances the stimulated emission.

A simple laser structure like this is limited to pulsed operation and requires careful cooling because of the high threshold current density for laser action, up to $1,000 \, \mathrm{A \, mm^{-2}}$ at room temperature. A good deal of effort has gone into reducing this to a more reasonable $0.5 \, \mathrm{A \, mm^{-2}}$ in heterojunction laser designs, which are single crystal pn junctions between dissimilar semiconductors. The crystallinity requirement means that the two materials must have a small lattice mismatch to enable the crystal to form continuously through the junction region. For example, semiconductors of composition $Al_x Ga_{1-x} As$ (Al-gas) can be grown on a GaAs substrate, to make devices of the kind shown in Figure 8.18. The band diagram has a stepped structure, corresponding to the change in band gap through the device. This makes an additional barrier to electron injection and confines the population inversion zone to a very thin central region. Further, the light created by recombination tends also to be confined to that region by total internal reflection, because the refractive index depends on the composition. These effects combine to make it easier to create both a population inversion and a high radiation density, so the device does not need to be forward biased so hard before laser light is emitted.

Communication using lasers is accomplished by switching the current above and below the threshold to make pulses of light. Switching frequencies above 2 GHz have been achieved, limited mainly by a turn on delay between the application of current and laser emission. Spectral purity is important to long distance transmission of light through a dispersive dielectric such as glass. The refractive index rises at shorter wavelengths, and the velocity of light within the medium falls. A light pulse containing different wavelengths may start its journey sharp, as in Figure 8.19a, but after transmission the shorter wavelength components will be lagging, as in Figure 8.19b. This is called smearing, and clearly limits the maximum pulse repetition rate over a given transmission distance. One solution is to make a highly tuned laser cavity by including an external diffraction grating within the resonant structure. This can reduce the bandwidth of the laser light by several orders of magnitude.

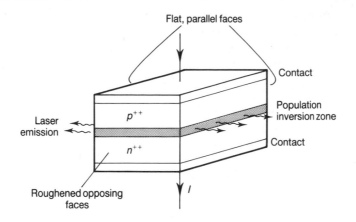

Figure 8.17 Simple junction laser structure

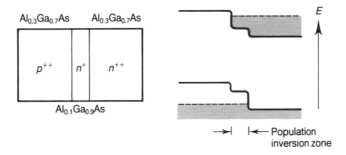

Figure 8.18 Heterojunction laser structure and band diagram

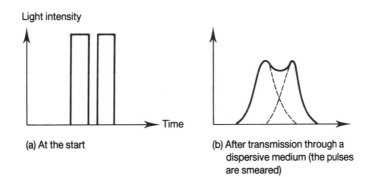

(a) At the start

(b) After transmission through a
dispersive medium (the pulses
are smeared)

Figure 8.19 Light intensity against time for pulses composed of many wavelengths

8.9 Optical fibres

An optical fibre is a thin filament of glass along which light may propagate. The thread is fashioned with a core of higher refractive index than that of the outer cladding, so that light rays undergo total internal reflection at the interface (which may be sharply stepped or graded); see Figure 8.20. The histories of the laser and the optical fibre are closely intertwined; the first lasers were cumbersome affairs, using a ruby rod or a gaseous lasing medium. They nevertheless demonstrated the potential for optical communication. The beams of light were intense and among the most parallel yet produced, but the divergence was still sufficiently great to expand the beam from millimetres to several metres in diameter after a few kilometres propagation. The beam path was also unstable, being sensitive both to mechanical vibration at the source and to variations in the refractive index of the air with temperature.

The principle of the light pipe had been known for decades, but it was restricted to ornamental use such as the internal illumination of water fountains. The refractive index of water is higher than that of air, so light shining within the water is ducted along the fountain jets, with enough escaping to make the display visible. In 1970 the technology of spinning glass fibres with the necessary internal structure was demonstrated. A good deal of the work was in making glass of sufficient perfection and purity: ordinary window glass will stop 99% of transmitted light within a few metres, while present day fibres display a minimum attenuation of 0.2 dB/kilometre.

With this development the propagation medium was ahead of the light sources, which were large and limited in their switching ability, until the production of the semiconductor junction laser. Coupling the light into the fibre requires some ingenuity, and may be accomplished by bonding the LED or laser to the fibre, either directly or via a small lens.

A further advance was the creation of optical fibres with an extremely narrow core, called monomode fibres to distinguish them from the earlier multimode types. If the core is broad there are several possible ray paths (modes) along the fibre, each with different lengths (Figure 8.21a). This leads to smearing between adjacent pulses when long-path light from one pulse arrives at the same time as short-path light from the next pulse. Monomode fibre supports

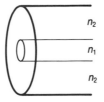

Figure 8.20 Optical fibre with stepped refractive index $n_1 < n_2$

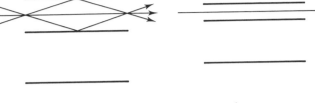

(a) Multimode   (b) Monomode

Figure 8.21 Propagation paths in optical fibres

only one mode of propagation (Figure 8.21b) and thus opens the way to even higher pulse frequencies.

At present the first undersea optical fibre cables are being laid. Despite appearances the fibres are stronger than copper wires of equivalent size. Fibres can carry far more information than conventional cables, and, unlike cables, are virtually free from cross-talk between adjacent lines. The information transmitted is virtually immune from electromagnetic noise and is far more secure against tapping. On the debit side, coupling light in and out of fibres is more difficult, as is joining fibres together. However, the stage is set for a rapid growth in optical fibre communications.

8.10 Liquid crystal displays (LCDs)

As the name suggests, liquid crystals are liquid materials which can possess a high degree of order. The molecules are shaped like long cylinders which readily align with one another because they are polar and are very responsive to electric fields. Three classifications are recognized:

1. *Smectic*: the molecules are aligned along their long axes and are in layers.
2. *Nematic*: the long axes are aligned without forming layers.
3. *Cholesteric*: the molecules form sheets which are slightly rotated with respect to its neighbours.

Nematic materials are most widely used in liquid crystal displays (LCDs). They are optically active, which means that the axes of polarized light are rotated by passage through the liquid crystal. The optical activity is strongly anisotropic (direction dependent), occuring most for light travelling perpendicular to the long molecular axes. This is illustrated in Figure 8.22.

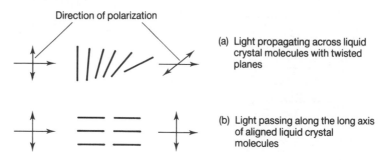

Figure 8.22 Rotation of the plane of polarization in liquid crystals

LCDs are constructed as in Figure (8.23) to make use of these phenomena. Light enters the device through a polarizing filter which only transmits light close to a given axis of polarization. It passes through the liquid crystal and encounters a second polarizing filter at right angles to the first. Behind this is a reflecting surface which causes the light to retrace its path. The surfaces in contact with the liquid crystal are textured (by abrasion or by evaporating material at an angle) in such a way that the liquid crystal molecules tend to align with the surface. The opposing surfaces are oriented perpendicularly, so that the molecular axes twist through 90° across the gap. Polarized light entering the crystal is *guided* by this twist, so that the light will also pass through the second polariser and back through the first.

If an electric field is set up across the crystal by applying a potential to transparent electrodes on either side the molecules swing round to align their long axes with the stronger field. In this orientation the molecules have almost no effect on polarized light. Any light passing through the first polarizer will be stopped by the second polarizer and will not be reflected back out. This makes the liquid crystal look dark because all the light falling on it is absorbed by one of the polarizing filters. When the electric field is removed the molecules swing back to their original orientation and the liquid crystal appears light again, in about 0.1 seconds.

A complete display has many electrodes which can be energized separately to generate patterns of dark shapes on a light background. The seven-segment configuration is widely used (Figure 8.24) and can display all decimal digits and a selection of upper and lower case characters. (The reader might care to work out which letters can be displayed.) An array of at least 7×5 dots can be used to display all conventional alphanumeric characters. In practice a square-wave alternating potential of 4–17 V_{rms} is applied, because a dc driving potential eventually dissociates the liquid crystal. LCDs are preferred above LED displays in battery-driven equipment because an LCD load is essentially capacitive, drawing only a few microwatts power. The disadvantage of LCDs are the need for external illumination and a more restricted temperature range.

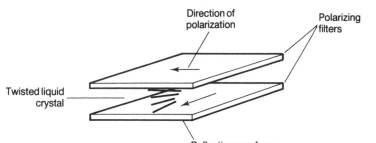

Figue 8.23 LCD construction

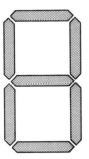

Figure 8.24 Seven-segment display

LCDs are beginning to be made with a large enough area and sufficiently small individual electrodes to be used in place of cathode ray tubes for televisions. Schemes have also been proposed to render colour images using LCDs, so there is a real possibility of banishing cumbersome television boxes over the next twenty years or so. Defect-free manufacture of fast-switchable LCD pixels is not trivial but appears feasible.

Summary

Main symbols introduced

Symbol	Unit	Description
A_{21}	s^{-1}	Einstein coefficient for spontaneous emission
B_{12}	$J\,m^{-3}$	Einstein coefficient for photon absorption
B_{21}	$J\,m^{-3}$	Einstein coefficient for stimulated emission

Symbol	Unit	Description
e		emissivity of a surface
g_{ph}		gain of a photodetector
I_{ph}	A	photocurrent
$R(v)$	$J\,s\,m^{-3}$	radiation energy density per unit frequency interval
α	m^{-1}	light absorption per unit length
η		quantum efficiency of a photodetector
σ_{SB}	$W\,m^{-2}\,K^{-4}$	Stephan–Boltzmann constant 5.67×10^{-8}

Main equations

$$F = F_0 \exp -\alpha x \text{ (photon flux decay by absorption)} \qquad (8.5)$$

$$I = I_0\left(\exp\left(\frac{qV_A}{kT}\right) - 1\right) - I_{ph} \text{ (photodiode characteristic)} \qquad (8.14)$$

Exercises

1. Find the photon flux and the luminous flux for 50 mW of light energy from red, yellow and blue LEDs emitting at 635, 585 and 530 nm respectively. Use the eye response curve given in Figure 8.4.

2. A certain PIN diode absorbs 98% of the photons which enter its intrinsic region from the p-type side. Determine the maximum thickness of the p-type side and the minimum thickness of the intrinsic region if 95% of the light entering the device is to be absorbed in the central region. Take $\alpha = 10^5$ m^{-1}. How will minority carrier diffusion influence these figures?

3. A PIN photodiode of area 1 cm^2 reflects 10% of the incident light and 97% of the remainder is absorbed within the depletion region. Find the photocurrent which flows when it is illuminated with 600 nm wavelength light of power density 80 W m^{-2}, assuming a quantum efficiency of one.

4. Find the photocurrent in a 10 mm × 10 mm photoconductor illuminated with the same light as in (3) if $\mu_e = 0.4$, $\mu_h = 0$, $\tau_e = 20$ ms, $\eta = 1$ and 10 V is applied to the device. Assume all the light is usefully absorbed. Find the dark current if the equilibrium electron number density is 10^{18} m^{-3} and the device is 0.1 mm thick.

5. The absorption coefficient in a certain photoconductor is 2.10^5 m^{-1}, and its dimensions are 5 × 5 mm, with 15 V applied across two opposite edges. It is illuminated with a photon flux of 3.10^{18} m^{-2} s^{-1}, with a quantum efficiency of 0.9. Find the thickness of the device if the dark current is 5% of the photocurrent under these conditions. Take $\bar{n} = 10^{19}$ m^{-3}, $\mu_e = 0.2$, $\tau_e = 50$ ms.

6.   A solar cell has a dark reverse bias current of 2 nA and an illuminated short circuit current of 40 mA. Find the open circuit voltage (a) in the dark; (b) when illuminated. Take $kT/q = 26$ mV.

7.   The illuminated solar cell of (6) is connected across a load resistance R. Find the power transferred to the load if $R = 100\ \Omega$ and $R = 10\ \Omega$.

8.   To use solar cells as a true alternative energy source, each cell must produce at least enough energy to fabricate its replacement during its working life. Find the minimum life of a cell to achieve this if the efficiency is 10% and the average insolation is 1300 W m^{-2} for seven hours/day. Assume the cell, of thickness 0.5 mm, is heated from room temperature to 1,000 °C ten times during manufacture. Take the specific heat to be 700 J kg^{-1} K^{-1} and the density to be 2,300 kg m^{-3}.

9.   Another way of referring to a population inversion is as a negative temperature. By applying the Boltzmann relation to a population inversion show how this concept arises. (The use of temperature in a system so far from equilibrium is of course a nonsense.)

10. Draw band diagrams to show that population inversion in a *pn* junction laser requires that the *n*-side Fermi energy is above the *p*-side conduction band edge and the *p*-side Fermi energy is below the *n*-side valence band edge.

9 | Metal–semiconductor junctions

9.1 Metal–metal junctions

Until now we have assumed that wires can be joined freely to semiconductors, without any thought as to the nature of the contact. In this chapter we will find that two broad types of metal–semiconductor contact are possible:

1.   *Ohmic contacts,* which possess only a small resistance and can pass current in either direction.
2.   *Schottky barrier diodes,* which pass current in one direction only.

Junctions between dissimilar metals will be considered as an introduction. The band diagram of a metal was given in Chapter 3 and is repeated here in more detail (Figure 9.1). There is no band gap. Most states below the Fermi energy are filled and most above E_F are empty. Those electrons within a few kT of the Fermi level will have empty states close to hand and will be free to move in response to an applied field. Typically one electron per atom is in this condition, which leads to a mobile electron number density of order 10^{28} m^{-3}, at least 1,000 times higher than the most heavily doped semiconductor.

The *work function* ϕ is the amount of energy, in electron-volts, between E_F and the vacuum energy level, E_{vac}. It is the amount of energy required to liberate an electron from the solid, by thermal excitation, photon absorption or

158

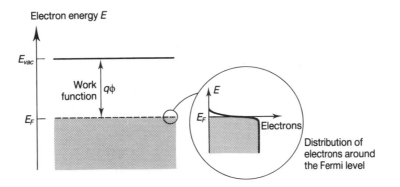

Figure 9.1 Band diagram of a metal

particle bombardment. (Strictly, ϕ varies a little according to the means used to free an electron, but we will ignore this.) The work functions of a few common substances are given in Table 9.1. Most are in the range 4–5 eV. Thorium is often used in conjunction with tungsten to increase the electron emission from heated cathodes, because of its lower work function.

To draw the band diagram of a junction between two metals with work functions ϕ_1 and ϕ_2 ($\phi_1 > \phi_2$) we follow the steps outlined in Chapter 4.

1. Draw the band diagrams of the two materials side by side with the vacuum levels aligned (Figure 9.2a). An electron just released from metal 1 has the same energy as one just released from metal 2, so $E_{vac1} = E_{vac2}$. Imagine now the situation of an electron close to the Fermi level in metal 2 as the materials are brought into firm contact. It sees in metal 1 many empty levels with a lower energy. Not only that, but the levels which it may

Table 9.1 Work functions of common substances

Material	Work function (eV)
Aluminium	4.1
Caesium	1.9
Carbon	4.8
Copper	4.3
Gold	4.8
Iron	4.6
Platinum	6.3
Silver	4.7
Thorium	3.5
Tungsten	4.5

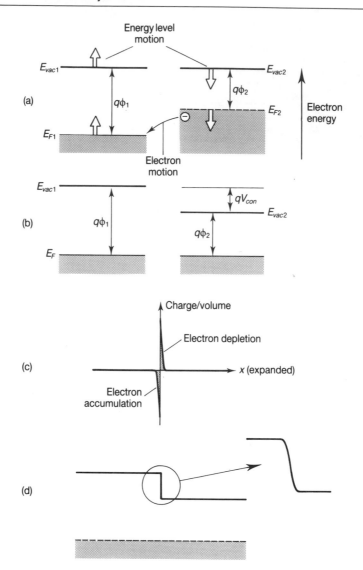

Figure 9.2 Formation of a metal–metal junction

legitimately occupy, according to the Fermi–Dirac distribution, are way down at E_{F1}. Random agitation causes it to jump and it stays there because of its lower energy condition. In moving across, the electron has transferred negative charge from metal 2 to metal 1. Instead of both being neutral, metal 1 is negatively charged and metal 2 is positively charged. This means that an electron released from metal 1 into vacuum will have a higher energy than one released from metal 2, so E_{vac1} becomes greater than E_{vac2}, and the difference in Fermi levels diminishes a little. More and more electrons cross, charging metal 1 more and more negative until the Fermi levels align and the process stops. It is important to grasp this chain of events now, because a similar argument will be used to derive the more complex metal–semiconductor band diagrams later on.

2.   Now draw a constant Fermi level and put the two vacuum levels above it (Figure 9.2b). Contact does not alter the work junction of each side, so there is a *contact potential difference* V_{con} formed at the junction:

$$|V_{con}| = |\phi_1 - \phi_2| \text{ V} \qquad (9.1)$$

3.   Finally let us consider the detail of the junction region. Metal 1 has gained electrons and metal 2 has lost them. Electrostatic attraction will keep the charges close to the junction. They take the form of very narrow sheets of charges, because of the large number of available states in both metals (Figure 9.2c). This is in contrast to a semiconductor where the charge per volume in depletion is limited to the doping level.

      These narrow, highly charged regions translate into very tightly curving energy levels, curving upwards on the left in metal 1 and downwards in metal 2. We will draw it as a sharp transition, but the reader must remember that a charge depletion and a charge accumulation region exists on either side, as in the expanded detail (Figure 9.2c).

What about the conduction properties of such a junction? There is no potential barrier at the Fermi energy, where the mobile electrons reside. If an external potential is applied to the junction it will set up an electric field in each metal which will drive electrons towards the more positive potential, as in Figure 9.3. The sloping vacuum level indicates the strength of the electric field, while

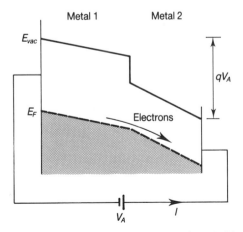

Figure 9.3 Voltage V_A applied to a metal–metal junction

the Fermi level shows that a current flows. A greater slope has been given to metal 2, supposing it to be more resistive and therefore to take up more of the applied potential.

9.2 Thermocouples

This section may be skipped without losing continuity. It is included because the student is often left with the impression that temperature measurement using metal junctions has something to do with the contact potential. This is quite wrong. Another physical phenomenon is responsible for producing a potential difference which varies with temperature: the Seebeck effect.

It was discovered that a temperature gradient in a conductor causes a potential difference between its ends. The Seebeck coefficient α_s is defined by:

$$\alpha_s = \frac{dV}{dT} \, V \, K^{-1} \tag{9.2}$$

so that a temperature difference ΔT along a conductor produces a potential difference $\alpha_s \Delta T$ between its ends. The potential originates in the differences in random motion of charges at each end, and may take opposite signs in different materials.

In order to sense this potential difference a circuit must be formed. If the same material is used on each side of the voltmeter, the Seebeck potentials in each arm are equal and opposing, giving a zero reading (Figure 9.4a). Using materials with different Seebeck coefficients in each arm overcomes this difficulty, and the contact potentials around the circuit cancel (Figure 9.4b; see exercise 1). This arrangement does require that the voltage measuring device is at a uniform, known temperature.

The best method is to place the voltmeter in one arm of the thermocouple, where it will again sense the difference in Seebeck potentials of the two materials. One junction is put at the temperature to be sensed, and the other is put at a different, defined temperature – frequently a 0 °C ice/water mixture (Figure 9.4c). It does not matter where the potential gradients occur in the wires because the measured potential is the integrated sum of all the voltage gradient along them.

It is common to use pairs of materials with Seebeck coefficients of opposite sign, so as to produce a reinforcing voltage for ease of measurement. Alloys of chromium and aluminium with nickel called chromel and alumel are such a pair often used for measurements up to a few hundred degrees Celsius. Thermocouples using different proportions of platinum and rhodium (Pt + 10% Rh: Pt + 13% Rh) are used up to 1,400 °C. Gold–iron couples can be used to sense very low temperatures. In practice the Seebeck coefficient is not constant with temperature and tables must be used to convert from measured potential to temperature for a given pair of materials.

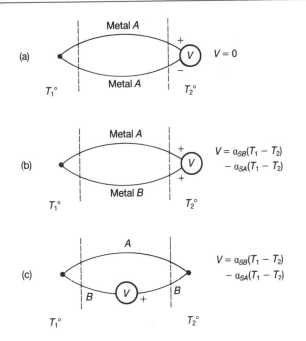

Figure 9.4 Thermocouple temperature sensing

9.3 Schottky barrier diodes

The Fermi level in a semiconductor depends on the doping, but the work function is defined in the same way as for a metal: it is the separation between the Fermi level and the vacuum level, in electron-volts. An additional quantity called the electron affinity is defined in semiconductors. It is the width of the conduction band in electron-volts and is given the symbol χ:

$$q\chi = E_{vac} - E_c \text{ J} \tag{9.3}$$

The distinction is drawn out in Figure 9.5. The electron affinity has a physical meaning, being the minimum amount of energy required to free an electron from the semiconductor. Remember that even in a p-type semiconductor there are a few electrons in the conduction band. They reside in its lowest energy levels, and so only need χ eV more energy to release them.

Consider now a junction between a metal of work function ϕ_m and an n-type semiconductor of work function ϕ_s, with $\phi_m > \phi_s$. The band diagram just before contact is given in Figure 9.6a, with the vacuum levels aligned. There are two charge transfers to be considered.

1. Electrons in the semiconductor conduction band move down to the Fermi level in the metal.

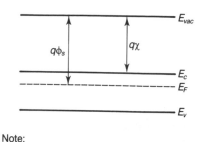

Note:
Only the work function depends on the doping.

Figure 9.5 Work function ϕ_s and electron affinity χ of a semiconductor

2. Electrons in the metal move down to the vacant valence band levels in the semiconductor. (Alternatively, we can say that valence band holes 'bubble up' to the metal Fermi level.)

In an n-type semiconductor there are far more electrons than holes, so the first charge movement dominates. The metal charges negative while the semiconductor becomes partly depleted of conduction band electrons, causing the Fermi levels to align (Figure 9.6b). The charge distribution in the junction is shown in Figure 9.6c: a thin electron accumulation region in the metal is balanced by a wider depletion region in the semiconductor.

Electrons in the metal and in the semiconductor conduction band both face potential barriers preventing more of them crossing. Conduction band electrons see a hill of height $(\phi_m - \phi_s)\,\text{eV}$, while those in the metal face a barrier of height V_b:

$$V_b = \phi_m - \chi\,\text{V} \quad (n\text{-type}) \tag{9.4}$$

A few energetic electrons do cross over from the conduction band into the metal to compensate for the few holes which will diffuse up the favourable potential gradient into the metal.

What happens when a potential is applied across this junction? We recall a result from the biasing of pn junctions: *most of the potential is dropped across the more lightly doped side*. The semiconductor has a far lower value of charge/volume in depletion than the metal does in accumulation. Look at Figure 9.7. An applied potential which tends to reduce the contact potential will therefore mostly serve to reduce the voltage dropped across the semiconductor depletion region, which in turn reduces its width. Likewise an applied potential which reinforces the contact potential will widen the depletion region. Now consider what happens to the barriers to electron movement. In the first case (Figure 9.7a), the barrier preventing conduction band electrons from entering the metal is reduced by qV_A. Electrons can cross over in large numbers and flow towards the positive plate of the battery. In the second case (Figure 9.7b)

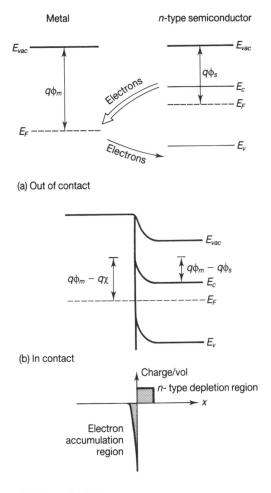

Metal   *n*-type semiconductor

(a) Out of contact

(b) In contact

(c) Charge distribution

Figure 9.6 Metal–*n*-type semiconductor junction $\phi_m > \phi_s$

electrons are drawn from the metal towards the semiconductor, but the barrier height V_b is virtually unchanged because most of the potential is dropped across the semiconductor depletion region. This is reverse bias and very little current flows.

This junction possesses rectifying properties and is called a Schottky barrier diode. Another such diode is formed between a metal and a *p*-type semiconductor, if $\phi_m < \phi_s$. The band diagram is derived in the same way, by drawing the materials just out of contact with the vacuum energies aligned (Figure 9.8a). Electrons will again tend to transfer from the semiconductor conduction band

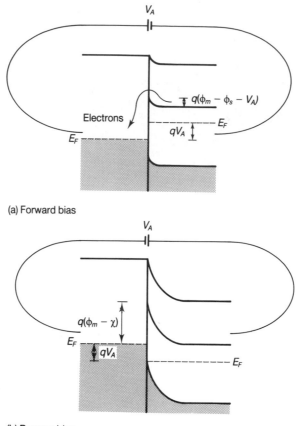

(a) Forward bias

(b) Reverse bias

Figure 9.7 Applying potential across a metal–n-type semiconductor junction $\phi_m > \phi_s$

into the metal, but these are only minority carriers in p-type material. The movement of electrons from the metal into the vacant valence band states charges the semiconductor negative and causes the Fermi levels to align (Figure 9.8b). Once again there is a depletion region in the semiconductor, due this time to the loss of holes near the interface. Holes in the semiconductor face a barrier of height $q(\phi_s - \phi_m)$ preventing them crossing into the metal. Vacant states in the metal only exist near and above E_F, and face an even larger barrier of

$$qV_b = E_g + q\chi - q\phi_m \text{ J} \quad (p\text{-type}) \tag{9.5}$$

preventing their transfer into the semiconductor. The few holes which do cross from the semiconductor balance the flow of minority electrons. The charge distribution at the junction is shown in Figure 9.8c.

If bias is applied to this junction most of the voltage is dropped across the

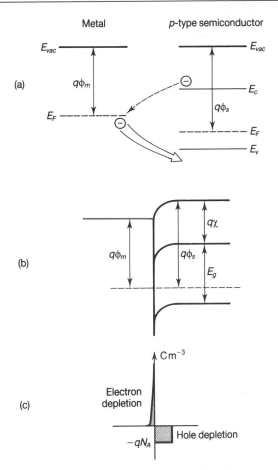

Figure 9.8 Metal–*p*-type semiconductor junction $\phi_m < \phi_s$

less densely charged depletion region, modifying the barrier seen by the holes in the valence band:

1. *Metal negative*: barrier is reduced, and holes flow into the metal.
2. *Metal positive*: barrier is raised and no current flows.

Schottky barrier diodes have a lower turn on voltage than *pn* junctions in forward bias, needing only 0.2–0.3 V while *pn* junctions require 0.6–0.7 V. They can be used as voltage clamps, especially in parallel with *pn* junctions (Figure 9.9). With a suitable choice of metal, such as aluminium with a silicon semiconductor, it is possible for just one of the metal–semiconductor junctions to be a Schottky barrier diode, while the other is an ohmic contact. The metal strip does not short out the *pn* diode but creates a parallel circuit as shown.

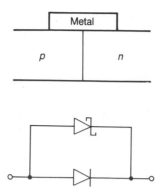

Figure 9.9 A Schottky-clamped *pn* junction

When the pair is forward biased the Schottky diode saturates first and prevents the build up of a large injected minority carrier charge in the *pn* junction. This means that the *pn* junction switch off time can be reduced considerably.

The turn off time of a Schottky barrier diode is very short, just a few nanoseconds. This is because carriers injected over a reduced barrier in forward bias quickly fall to the Fermi energy level in the metal (Figure 9.10). If the bias is reversed these carriers face a large barrier preventing them recrossing, in contrast to the *pn* junction where injected minority carriers can cross back to their home side, delaying the turn off of the diode.

Barrier heights at the surface of real semiconductors are different from that predicted from bulk measurements. This is because the break in the crystal introduces a large number of extra allowed levels throughout the band gap which are called surface states. Those states below the Fermi level are occupied while those above are mostly empty (Figure 9.11). The states are numerous and

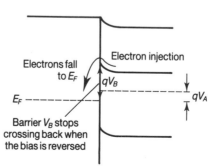

Figure 9.10 Injected electrons quickly fall to E_F in the metal

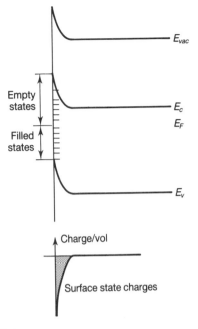

Note:
Barrier heights are altered by charged surface states.

Figue 9.11 Real semiconductor surface

act like a thin layer of heavy doping: a small change in the position of E_F causes a large number of surface states to fill or empty. Their effect, therefore, is to pin E_F close to a fixed value, about one-third of the way up the band gap.

9.4 Ohmic contacts

The most beautiful solid state devices would be quite useless if all metal–semiconductor junctions were diodes, because it would be impossible to interconnect them with metallic conductors. Junctions which obey Ohm's law and allow current to flow in either direction are called ohmic contacts. Naturally, the contact resistance should be as low as possible, just a few milliohms. Here we will consider the band diagrams of ideal ohmic contacts and will use them to demonstrate that their I/V characteristics are symmetrical and linear.

9.4.1 Metal: n-type, $\phi_m < \phi_s$

The band diagram for the materials just out of contact is drawn in Figure 9.12a, and the equilibrium band diagram with the Fermi levels aligned is given in

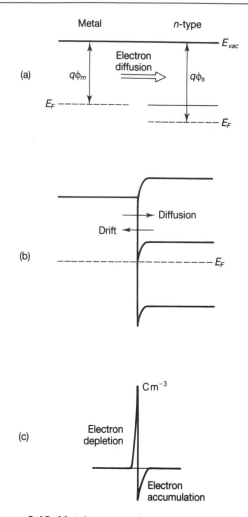

Figure 9.12 Metal–n-type ohmic contact, $\phi_m < \phi_s$

Figure 9.12b. The loss of electrons from the metal into the semiconductor is brought about by diffusion in this case. Although conduction band electrons find it energetically favourable to enter the metal they are heavily outnumbered by electrons diffusing in the opposite direction until the contact potential difference is established. The charge distribution in Figure 9.12c indicates that the negatively charged region in the semiconductor (shown by the upward curvature of the band diagram) is due to an accumulation of electrons at the interface. As the conduction band edge dips towards the Fermi level it creates a low energy pocket where electrons collect, or else we can say that the material becomes increasingly n-type in this region.

What is the I/V characteristic of this junction? If the metal is biased positive with respect to the semiconductor, there is no barrier at all to electron flow. The band diagram consists of sloped bands and a sloped Fermi level (Figure 9.13a) corresponding to current flow, with most of the potential dropped across the more resistive semiconductor. Electrons in the metal face a potential barrier limiting the number diffusing into the conduction band. A positive bias applied to the semiconductor upsets the dynamic equilibrium by lowering the barrier a little and allowing electrons to flood across. Current flows and once again most of the voltage is used to drive current through the resistance of the semiconductor (Figure 9.13b).

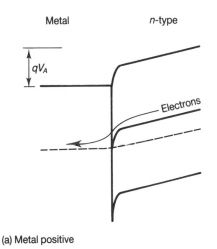

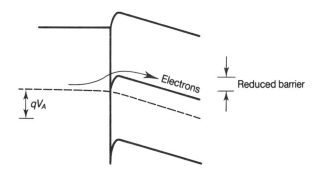

Figure 9.13 Band diagrams for a biased ohmic contact

9.4.2 Metal: p-type, $\phi_m > \phi_s$

Figure 9.14 details the formation of this contact. When the Fermi energy in the metal is below the valence band edge E_v, valence band electrons find it energetically favourable to enter the metal, or in terms of hole transport, holes move from the metal into the valence band. To complete the alignment of the two Fermi levels requires more holes to flow from the metal into the semicon-

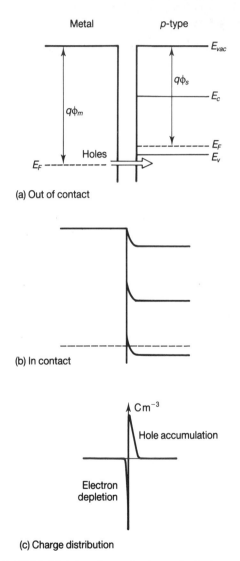

Figure 9.14 Metal–p-type ohmic contact, $\phi_m > \phi_s$

ductor. The large number of holes near to the Fermi level in the metal provides a concentration gradient strong enough for holes to diffuse into the valence band, even when hole energy is increased by doing this. The process is similar to the formation of a *pn* junction, where diffusion is held in check by a potential barrier. Again there is an accumulation of majority carriers within the semiconductor. Biasing the metal negative draws holes directly into the metal, and a positive bias reduces the barrier for hole diffusion from the metal into the semiconductor. This contact is also non-rectifying.

9.4.3 Heavily doped ohmic contacts

If the semiconductor surface is very heavily doped a nominally rectifying metal–semiconductor junction can become an ohmic contact. This is because the depletion region is made so thin that electrons can tunnel through the barrier rather than requiring the energy to surmount it. Practical contacts are not sharp since a little of the metal penetrates the semiconductor. Thus gold containing a little tin can form an ohmic contact with *n*-type silicon, although $\phi_m > \phi_s$ in this case. Tin is a donor dopant in silicon, and can increase N_D at the surface enough to make the depletion region barrier transparent to electrons. Similarly aluminium is a p-type dopant and makes a good ohmic contact to p-type silicon, although $\phi_m < \phi_s$. If, as is often the case, it is necessary for an aluminium film to make ohmic contacts to both *p*- and *n*-type silicon, it is necessary to dope the *n*-type region heavily beforehand, so the semiconductor surface remains *n*-type when the metal enters the surface. By doping parts of the surface heavily and parts lightly before depositing the metal film, both ohmic contacts and Schottky barrier diodes can be formed simultaneously.

EXAMPLE 1

■ What kind of contact is an ideal junction between tungsten and silicon doped with 10^{23} m^{-3} acceptors? Use the semiconductor data given in Table 3.1.

☐ The work function of tungsten is 4.5 eV (from Table 9.1). The Fermi level in the semiconductor is found using equations (3.20) and (3.22):

$$E_{Fi} = \frac{1}{2}(E_c + E_v) + \frac{kT}{2}\ln\frac{N_c}{N_v} \text{ J} \tag{3.20}$$

$$= \text{(midgap)} + \frac{1.38 \times 10^{-23} \times 300}{2}\ln 2.8$$

$$= 2.13 \times 10^{-21} \text{ J} \quad \text{above gap centre}$$

$$\text{or } 0.013 \text{ eV} \quad \text{above gap centre}$$

Then:

$$E_F = E_{Fi} + \frac{kT}{2}\ln\frac{n}{p}\,\text{J} \tag{3.22}$$

The material is p-type, with $p \simeq N_A$ and $n \simeq n_i^2/N_A$:

$$E_F - E_{Fi} = \frac{kT}{2}\ln\frac{n_i^2}{N_A^2}$$

$$= -6.5 \times 10^{20}\,\text{J}$$

$$\text{or } -0.406\,\text{eV}$$

So E_F is $0.406 - 0.013 = 0.393$ eV below the centre of the gap. The band gap is 1.12 eV, and the electron affinity is 4.05 eV, so the work function is:

$$\phi_s = 4.05 + \frac{1.12}{2} + 0.393$$

$$= 5.003\,\text{eV}$$

The semiconductor is p-type, and $\phi_s > \phi_m$, so an ideal abrupt junction would be a Schottky barrier diode.

EXAMPLE 2

■ What is the width of the semiconductor depletion region at zero bias for above contact?

☐ The contact potential is dropped almost entirely across the depletion region, as seen in the relevant band diagram. The metal acts like a heavily n-type region in this instance, so the formula for a single-sided pn junction can be used (from equation (6.51)):

$$d = \left(\frac{2\epsilon\text{V}}{qN_A}\right)^{1/2}$$

$$= \left(\frac{2 \times 11.9 \times 8.85 \times 10^{-12} \times (5.003 - 4.5)}{1.6 \times 10^{-19} \times 10^{23}}\right)^{1/2}$$

$$= 8.14 \times 10^{-8}\,\text{m}$$

EXAMPLE 3

■ Find the barrier heights preventing electrons crossing between gold and silicon in an ideal unbiased contact, if the Fermi level is 0.2 eV below the edge of the conduction band in the semiconductor.

☐ The work function in gold is 4.8 eV. The semiconductor work function is the sum of the electron affinity and $(E_c - E_F)/q$ eV: $\phi_s = 4.05 + 0.2 = 4.25$ eV. The semiconductor is n-type, because E_F is well above the centre of the band gap. ϕ_m is greater than ϕ_s, so the contact is a Schottky barrier diode. The barrier heights can be obtained from the band diagram, Figure 9.6b.

For electrons in the semiconductor:

Barrier $= \phi_m - \phi_s = 0.55$ eV

For electrons in the metal:

Barrier $= \phi_m - \chi = 0.75$ eV

Summary

Main symbols introduced

Symbol	Unit	Description
E_{vac}	J	vacuum energy level
V_b	V	Schottky barrier potential
V_{con}	V	contact potential
α_s	V K^{-1}	Seebeck coefficient
ϕ	eV	work function
χ	eV	electron affinity in a semiconductor

Main equations

$$V_{con} = |\phi_1 - \phi_2| \text{ V} \tag{9.1}$$

$$q\chi = E_{vac} - E_c \text{ J} \tag{9.3}$$

$$V_b = \phi_m - \chi \text{ V} \quad (n\text{-type Schottky barrier diode}) \tag{9.4}$$

$$V_b = E_g/q + \chi - \phi_m \text{ V} \quad (p\text{-type Schottky barrier diode}) \tag{9.5}$$

For ideal contacts:

	$\phi_m > \phi_s$	$\phi_m < \phi_s$
n-type	Schottky barrier diode	ohmic
p-type	ohmic	Schottky barrier diode

Exercises

1. Draw an equilibrium band diagram for a multiple junction of gold, copper, aluminium and gold. What is the contact potential difference between each end? Use the work function data in Table 9.1.

2. A copper: constantan thermocouple with a cold junction at 0 °C produces the following voltages at different temperatures:

T °C	50	100	150
mV	2.00	4.22	6.64

    Estimate the Seebeck coefficient for constantan at 75° and 125° if that of copper is zero. How would these results be affected if the contact potential difference were doubled?

3. Find the work function in silicon at 300 K (a) doped with 10^{23} m^{-3} donors; (b) doped with 10^{22} m^{-3} acceptors. The electron affinity is 4.05 eV, the band gap is 1.12 eV, $N_c = 2.8 \times 10^{25}$ m^{-3}, $N_v = 1.0 \times 10^{25}$ m^{-3}, $n_i = 1.5 \times 10^{16}$ m^{-3}. What is the minimum photon energy required to free (i) a few electrons (ii) a large number of electrons by photoemission?

4. Junctions are made with the semiconductor materials in (3), using (a) gold (b) aluminium (four junctions in all). Draw the ideal band diagram for each contact in equilibrium, and determine the type of I/V characteristic in each case. State clearly which polarity of bias is the forward bias in any rectifying junctions.

5. Find the depletion region width in a Schottky barrier diode reverse biased by 7 V if $\phi_m = 5.1$ eV, $\phi_s = 4.7$ eV, $N_D = 10^{20}$ m^{-3}, $\epsilon_r = 12$.

6. The contact in (5) can become ohmic if the depletion region thickness falls below 10 nm at zero bias. What value of doping is necessary to achieve this?

7. Draw band diagrams for an ideal metal: p-type semiconductor contact with the metal biased (a) positive; (b) negative with respect to the semiconductor, with $\phi_m > \phi_s$. Hence show that the contact does not rectify.

8. Sketch the band diagrams for ideal metal–semiconductor contacts if $\phi_m = \phi_s$, for both p-type and n-type material. Speculate on the I/V characteristics of such contacts.

9. The reverse saturation current and ideality factor for a Schottky barrier diode are typically 1 nA and 1.02 nA respectively, while the figures for a pn junction diode are 1 pA and 1.4 pA. Show that the semi-ideal forward bias voltage of the Schottky diode is less than that of the pn junction.

10. The I/V characteristic of a Schottky diode is given by:

$$I = CT^2 \exp \frac{-qV_b}{kT} \left[\exp \left(\frac{qV_A}{\eta kT} \right) - 1 \right]$$

The constant C for a particular diode is 3.10^{-6} A K^{-2}, the barrier height V_b is 0.5, and the ideality factor η is 1.04, all at 300 K. Find the reverse saturation current. Suggest a definition for the cut-in voltage and find its value in this case. How could V_b be measured?

10 Semiconductor device fabrication

10.1 Introduction

The properties and limitations of semiconductor devices are bound up with the methods used to manufacture them. Moreover, the semiconductor fabrication industry is growing in importance as circuits are increasingly designed to be made as a single integrated unit rather than from discrete packages each containing a single device or a few devices. In this chapter we shall review the main steps in the manufacture of semiconductor devices and attempt to identify the trends which will shape the industry over the next decades.

10.2 Semiconductor slices

We have emphasized that the active heart of solid state devices consists of pure, crystalline semiconductor possessing as few defects as possible. The highest purity metal normally available is 99.99995% pure, or about 1 part impurity in 10^6. This would be quite inadequate for semiconductor of device quality, which must contain no more than 1 part in 10^8 or even 1 part in 10^{10} impurity. Doping impurities must be added in a controlled fashion in quantities as low as 1 part in 10^9. Tight control is required at all stages of manufacture to create even one working device. Cleanliness is a necessary obsession of the industry.

Metallurgical grade silicon, about 99.5% pure, is extracted from quartzite SiO_2 rock in an electric arc furnace by reaction with carbon. The silicon is then reacted with chlorine at high temperature to form $SiCl_4$, which is a volatile liquid. It is separated from other volatile halide compounds, principally metals, by distillation and the reaction is reversed at an intermediate temperature, depositing electronic grade silicon (EGS) on a heated silicon rod. This material is pure but polycrystalline.

The dominant method for both silicon and GaAs crystal preparation is called the Czochralski process (say zo-kral-ski). The material is melted in a crucible and a small seed crystal is dipped into the melt. As it is slowly withdrawn, material clings to it and freezes with the *same* crystal pattern as the seed. (The origin of the first seed crystal is hidden in the mists of time.) The GaAs process is considerably more tricky because all the arsenic would boil away at the melting point of gallium unless it was prevented by an inert liquid capping layer under high inert gas pressure. Single crystals of silicon several metres long and weighing up to 50 kg have been prepared in this way. The crystal diameter can be 200 mm or more, with 100–150 mm being the most common range (in 1990).

A further purification technique called zone refining can be applied at this stage or to a cylinder of EGS. An induction heating coil is passed along the rod, melting the material closest to it. Surface tension forces stop the liquid from running away. The physics of solutions dictate that impurities will tend to segregate in the melt as it travels down the rod, much as sugar can dissolve more easily in hot tea than in ice. After several unidirectional passes, a high proportion of impurities are frozen at one end of the bar, called the tang end, which is sawn off and discarded.

The pure single crystal cylinder is machined to a uniform diameter and then sawn into thin discs, known as slices or wafers. At least one side is polished to a high degree of flatness. Slices are usually the end product of a dedicated factory and are the feedstock of other factories whose output is the finished device.

10.3 Lithography

Lithography means writing on stone; but it has come a long way from the hammer and chisel. The advances in optical photolithography (writing with light on stone) have been one of the main reasons for the rapid development of the semiconductor industry and it remains a cornerstone technology. The reason for this is highlighted in Figure 10.1, which summarizes the whole manufacturing sequence: while individual layer creation and etching processes are done once or twice, the patterning process happens six to twelve times as successive layers are added. Most of this description will concentrate on the central loop, but engineers should always bear in mind the considerable amount of inspection and testing work involved in solid state device manufacture, nor should they forget the last step for even a moment.

The photolithographic patterning process is outlined in Figure 10.2. Incoming wafers have been completely covered in a thin film, typically 1 micron thick, which may be a metal, dielectric or polycrystalline silicon. It is required to remove part of the layer and leave the rest, for example to provide interconnection paths. A typical process is as follows:

1.   A light-sensitive liquid called photoresist or resist is poured on the wafer. This spreads out into a thin uniform coating when the wafer is spun at a few thousand rpm on a vacuum chuck. The liquid is baked dry, turning the wafer into a photographic plate.

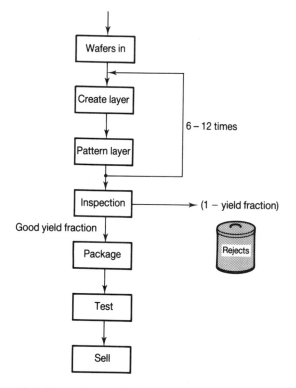

Figure 10.1 Flow diagram for integrated circuit manufacture

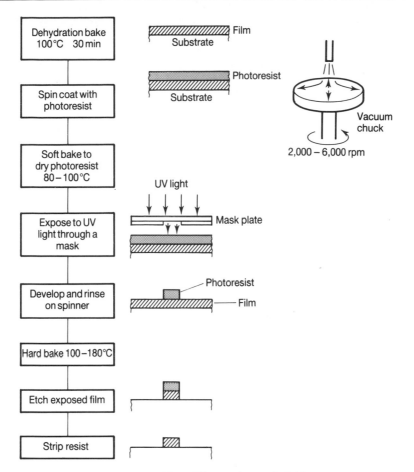

Figure 10.2 Photolithography and etching

2. The resist-coated wafer is exposed to ultraviolet light through a mask plate which contains the required pattern. Both positive and negative resists are available: light softens positive resist and hardens negative resist.

3. The exposed wafers are dipped or sprayed with a developer solution, which removes the softer parts of the resist. For a positive resist, the resist remaining has the same image as the light-blocking areas of the mask.

4. The wafer with the resist image is baked again at 70–200 °C to toughen the resist film. The purpose of the resist is to prevent the layer beneath it from being etched, so it must not be eroded significantly during the next step.

5. The wafer is etched to remove the unwanted, unprotected parts of the top layer.

6. The remaining resist is stripped, leaving the wafer ready for the next layer addition.

Wafers are often inspected before etching to check the quality of the resist image. It is comparatively simple to strip a bad resist layer and start the sequence again, but rework is almost out of the question if the wrong pattern has been etched into the underlying film.

10.4 Doping

10.4.1 Diffusion doping

Controlled amounts of doping impurities can be added to some semiconductors, notably silicon, by heating the wafer in a gaseous environment which contains the dopant. Usually a batch of 25–100 wafers are processed simultaneously in a quartz furnace tube. The concentration gradient which exists at the surface causes the doping atoms to diffuse into the material. Temperatures close to 1,000 °C are required because the diffusion coefficient, D, for impurities within a crystalline solid varies strongly with temperature:

$$D = D_0 \exp \frac{-E_a}{kT} \, \text{m}^2 \text{s}^{-1} \qquad (10.1)$$

D_0 is a constant and E_a is the activation energy for diffusion. The variation of dopant concentration $N(\text{m}^{-3})$ with depth x is of particular interest, because the doping profile $N(x)$ determines the position of the pn junctions. During diffusion doping the concentration is a function $N(x, t)$ of both position and time, which can be determined by considering the diffusion fluxes in and out of a small volume at position x of thickness dx (Figure 10.3).

$$\text{flux in at position } x = -D \, \frac{\partial N}{\partial x} \, \text{m}^{-2} \text{s}^{-1} \qquad (10.2)$$

$$\text{flux out at position } x + dx = -D \left(\frac{\partial N}{\partial x} + \frac{\partial^2 N}{\partial x^2} \, dx \right) \text{m}^{-2} \text{s}^{-1} \qquad (10.3)$$

Using $1 \, \text{m}^2$ area normal to the fluxes the rate of change of N within the volume $(1 \times 1 \times dx) \, \text{m}^3$ can be expressed in the same units: $(\partial N / \partial t) dx \, \text{m}^{-2} \text{s}^{-1}$. Hence:

$$\frac{\partial N}{\partial t} \, dx = -D \, \frac{\partial N}{\partial x} + D \left(\frac{\partial N}{\partial x} + \frac{\partial^2 N}{\partial x^2} \, dx \right) \text{m}^{-2} \text{s}^{-1} \qquad (10.4)$$

or:

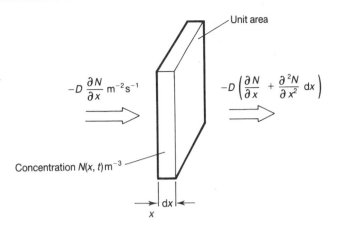

Note:
The number density of doping atoms in the slice changes with time t when the fluxes in and out of the slice are different.

Figure 10.3 Diffusion of dopants

$$\frac{\partial N}{\partial t} = D \, \frac{\partial^2 N}{\partial x^2} \text{ m}^{-3}\text{s}^{-1} \tag{10.5}$$

This is the diffusion equation, also known as Fick's second law. Partial differential equations are not easy to solve, but two cases are relevant to device fabrication.

1. *Predeposition*: this is a short, intense diffusion in which the surface concentration is held at the maximum the semiconductor can take, called the solid solubility limit. This limit also varies with temperature. The boundary conditions in this case are:

    $N(0) = N_0$ for all times t $\qquad\qquad\qquad$ (10.6)

    $N(\infty) = 0$ for all times t

The solution of the diffusion equation with these boundary conditions can only be given in terms of a new mathematical function. First we define the error function, given the symbol $\text{erf}(z)$:

$$\text{erf}(z) = \frac{2}{\sqrt{\pi}} \int_0^z \exp^{-y^2} dy \tag{10.7}$$

The curve $(2/\sqrt{\pi}) \exp -y^2$ is given in Figure 10.4. The total area under the curve is unity, and the area up to a point z is $\text{erf}(z)$. The remaining area from z to infinity is the complementary error function $\text{erfc}(z)$:

$$\text{erfc}(z) = 1 - \text{erf}(z) \tag{10.8}$$

The two new functions are shown in Figure 10.5 and $\text{erfc}(z)$ is tabulated in

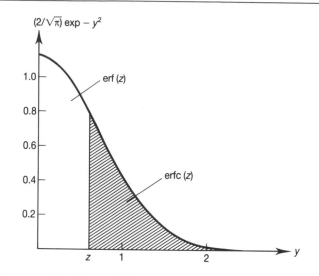

Figure 10.4 The error function erf(z) and the complementary error function erfc(z) expressed as areas under a curve

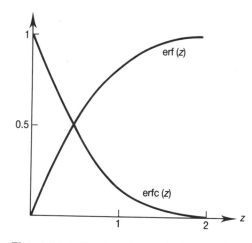

Figure 10.5 The functions erf(z) and erfc(z)

Table 10.1. Do not be alarmed by all these new terms: you have lived with $\sin(x)$ and $\cos(x)$ for years, and now here is one more function for your mathematical repertoire.

The predeposition profile is

$$N(x,\ t) = N_0 \operatorname{erfc}\left(\frac{x}{2\sqrt{Dt}}\right) \tag{10.9}$$

Since erfc$(0) = 1$, the surface concentration at $x = 0$ is always N_0, in

Table 10.1 The complementary error function erfc(z)

z	erfc(z)	z	erfc(z)
0	1.000	1.9	7.21×10^{-3}
0.1	0.888	2.0	4.68
0.2	0.777	2.1	2.98
0.3	0.671	2.2	1.86
0.4	0.572	2.3	1.14
0.5	0.479		
0.6	0.396		
0.7	0.322	2.4	6.89×10^{-4}
0.8	0.258	2.5	4.07
0.9	0.203	2.6	2.36
1.0	0.157	2.7	1.34
1.1	0.120		
		2.8	7.50×10^{-5}
1.2	8.97×10^{-2}	2.9	4.10
1.3	6.60	3.0	2.21
1.4	4.77	3.1	1.16
1.5	3.39		
1.6	2.36		
1.7	1.62	3.2	6.03×10^{-6}
1.8	1.09	3.3	3.06
		3.4	1.52
		3.5	7.43×10^{-7}
		3.6	3.56
		3.7	1.67

agreement with the boundary conditions. Likewise for large x the concentration is vanishingly small at all times. Note the reappearance of the quantity $\sqrt{(Dt)}$ in a new context. This is now the diffusion length for dopants at high temperature. They penetrate just a few diffusion lengths into the material because erfc(z) falls very quickly as z increases.

2. *Drive in*: a predeposition step can be followed by a drive in or push diffusion. No new impurities are added but those already present are redistributed. The dopants penetrate further into the bulk material and the concentration at the surface falls, preparing the way for the addition of more dopants, possibly of the opposite kind. The boundary conditions are:

$$\left. \frac{\partial N}{\partial x} \right|_{x=0} = 0 \quad \text{for all times } t \tag{10.10}$$

$$N(\infty) = 0 \quad \text{for all times } t$$

The first condition indicates that there is no diffusion flux at the surface, unlike predeposition. If the initial distribution can be approximated to a delta function, ie a large amount very close to the surface, the diffusion equation has the following solution:

$$N(x,\ t) = \frac{Q}{(\pi Dt)^{1/2}} \exp \frac{-x^2}{4Dt} \ \mathrm{m}^{-3}$$

(10.11)

This dies away quickly as x increases, but not so quickly as the complementary error function. Again the fall with distance is keyed to the diffusion length $(Dt)^{1/2}$, indicating that it penetrates further as time goes on. Q is the total amount of impurity per unit area of wafer surface, and remains constant with time. Equation (10.11) thus shows that the surface concentration falls with time.

With these processes we already have enough technology to create a *pn* junction diode:

1. Take a *p*-type semiconductor wafer. The doping profile is flat across all its depth (Figure 10.6a).
2. Diffuse a donor impurity by the two-step process given. The doping profile now contains two components (Figure 10.6b and c). In the surface regions on both sides $N_D > N_A$ because the whole wafer surface is diffused so the semiconductor there is *n*-type. The central region is *p*-type because $N_A > N_D$.
3. Grind or etch away the back surface of the wafer, leaving just one *pn* junction (Figure 10.6d). In future examples only the active front surface will be considered.

10.4.2 Ion implantation

Diffusion doping works well, but has three important limitations:

1. It is not precise, the total dose only being controlled to $\pm5\%$ or worse.
2. The highest concentration of dopants always occurs at the surface.
3. The total dose and the junction depth cannot be chosen independently.

Ion implantation is increasingly replacing diffusion doping because it scores better in all three areas, especially the first. It is a far more expensive technique on two counts: the equipment is more sophisticated and it processes only one wafer at a time, rather than a large batch. However, it is a must for compound semiconductors like GaAs, which cannot be heated above 700 °C with exposed surfaces because one component begins to boil away.

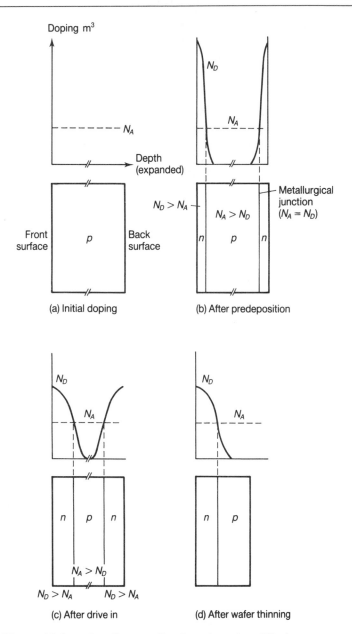

Figure 10.6 *pn* junction creation by a two-step diffusion process

In ion implantation positive ions of phosphorus, boron or other doping atoms are accelerated in a vacuum by falling through a potential of 3–500 kV and are fired at the semiconductor wafer. The ions become buried near the surface, the distribution often being close to a Gaussian curve (Figure 10.7):

$$N(x) = N_0 \exp \frac{-(x - R)^2}{2\sigma_R{}^2} \text{ m}^{-3} \tag{10.12}$$

where N_0 is the peak concentration, R is the depth of the peak below the surface (known as the range), and σ_R is the standard deviation of the distribution, which determines how far it spreads out. The range increases nearly linearly with ion energy, and is typically 0.1 microns for 100 keV ions. Far deeper penetration can result if channelling occurs, in which the ion beam finds an easy path through the crystal lattice. The dopant distribution is more irregular, as shown in Figure (10.7). Channelling can be used profitably in a few cases, but it is normally avoided by orienting the crystal off axis with respect to the incident beam.

One drawback of the ion implantation method is the immense damage done to the crystal lattice by the high energy ions. Also, the dopants themselves are not active unless they take the place of silicon atoms in a regular lattice. Fortunately, the crystal structure is a minimum energy configuration for the assembled atoms, which they will tend to take up provided they can move about with respect to each other. At room temperature such motion is virtually non-existent, but at several hundred degrees Celsius there is enough internal movement for the lattice to reform, and for dopant atoms to take up substitutional positions. After implantation, therefore, wafers are annealed at up to 1,000 °C to activate the dopants and repair the damage. This can even be done with compound semiconductors, provided an impervious capping layer such as silicon nitride is applied first. Unlike diffusion doping it is not necessary to etch windows in the capping layer, so no component of the compound is lost as a vapour.

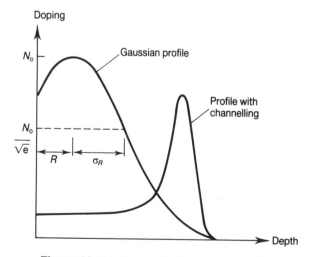

Figure 10.7 Ion implantation doping profiles

Other doping profiles can be created by summing successive ion implantations at different depths. For example, a region of uniform doping can be formed by two implants of similar doses and different ranges (sketch it and see).

10.5 Masking and silicon oxidation

The ability to stop dopants penetrating unwanted areas of the wafer is the second key to creating whole circuits on a single chip. The way this is done is to form a blocking or masking layer over the whole surface, and then to make openings or 'windows' in the layer by photolithography and etching, through which dopants can be introduced.

To be useful any masking layer must itself be able to withstand the conditions under which dopants are introduced and it must further stop the dopants getting into the semiconductor underneath. In the case of diffusion doping the high temperatures used limit the choice of material severely: silicon dioxide (SiO_2) and silicon nitride (Si_3N_4) are the only practical materials. Both of these have impurity diffusion coefficients much lower than those for silicon, and remain solid, adherent coatings at well over 1,000 °C.

Ion implantation occurs much closer to room temperature and sufficient thickness of almost any solid film will stop a beam of ions, so a wider variety of masking agents is available. Silicon dioxide or silica-based glasses can be used but so can the photoresist layer itself. This is preferable because it eliminates one layer creation step and one etching process, unless the layer is required in any case as a permanent part of the device structure.

Silicon dioxide is sufficiently important to merit a discussion to itself. (Device engineers call it oxide without ambiguity, so widespread is its use.) It is an insulator, with a band gap of 8 eV. Its coefficient of thermal expansion is $5 \times 10^{-7} \text{ K}^{-1}$, which is sufficiently close to that of silicon (2.6×10^{-6}) to prevent thermally induced stresses causing crystal defects in the underlying silicon in most circumstances. (Suppose a SiO_2 layer is formed at 1,000 °C on silicon. The silicon will try to contract more than the SiO_2 on cooling to room temperature, leaving the silicon in tension and the oxide film in compression.) It has four main roles in semiconductor devices:

1. As an insulator, separating two levels of conductors or semiconductors.
2. As a doping mask, especially for diffusion doping.
3. As a passivation layer, terminating the silicon crystal structure without leaving a large number of surface states.
4. Combining all these functions, it is an essential part of the insulated gate field effect transistor (IGFET or MOSFET: see Chapter 15).

Silicon dioxide films can be deposited but the best interface properties with silicon are obtained if the wafer itself is oxidized. Either oxygen or water vapour can be used to convert the surface layer of silicon to SiO_2:

$$Si + O_2 \longrightarrow SiO_2 \quad \text{(dry oxidation)} \tag{10.13}$$

$$Si + 2H_2O \longrightarrow SiO_2 + 2H_2 \quad \text{(wet oxidation)} \tag{10.14}$$

The reaction proceeds too slowly to be measured at room temperature, but films 250 nm–2,000 nm thick can be grown in a few hours using a furnace tube at 1,000–1,200 °C.

Unlike most deposition processes, the conversion rate of Si to SiO_2 is not constant with time. The oxidizing species has to diffuse through the growing SiO_2 layer and then react with silicon at the interface (Figure 10.8). Initially growth is controlled by the reaction, because it is easy to diffuse through a thin layer. In this case the film thickness d_{ox} does increase linearly with time:

$$d_{ox} \simeq \frac{B}{A}(t + \tau_0) \tag{10.15}$$

$B\tau_0/A$ is an initial oxide thickness, seen in dry oxidations. A and B are constants which depend strongly on temperature. After a while, it is the diffusion process which limits the rate of growth, which then slows to a square root dependence on time:

$$d_{ox} \simeq \sqrt{(Bt)} \text{ m} \tag{10.16}$$

These two relations are the limits for short and long oxidation times of the more general linear-parabolic law formulated by Deal and Grove. (B. E. Deal and A. S. Grove, 'General relationship for the thermal oxidation of silicon', *Journal of Applied Physics,* vol 86, 1965, p. 3770):

$$d_{ox} = \frac{A}{2}\left[\left(1 + \frac{t + \tau_0}{A^2/4B}\right)^{1/2} - 1\right] \text{ m} \tag{10.17}$$

A process can now be devised to make a *pn* junction diode at one or more places on a wafer, rather than over the whole wafer surface (Figure 10.9):

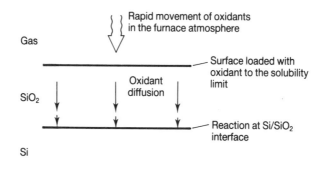

Figure 10.8 Processes in thermal oxidation of silicon

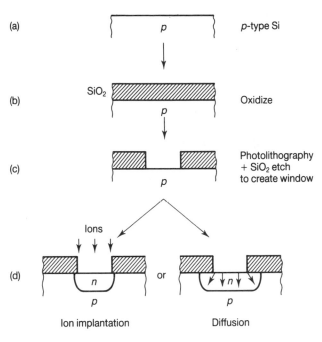

Note:
There is greater lateral spread of dopants in diffusion doping.

Figure 10.9 Doping through a silicon dioxide mask

1. Take a p-type wafer.
2. Grow a silicon dioxide masking layer over the whole surface.
3. Create windows in the layer where diodes are required.
4. Either
   (a) diffuse n-type dopants; or
   (b) implant n-type dopants and anneal the wafer.

10.6 Etching

The discussion on lithography assumed that it would be possible to etch through a thin film without harming either the underlayer or the material protected by the photoresist. Real etch processes are a compromise between the etch rates of the film, the underlayer and the resist itself, usually expressed as a selectivity $S_{A/B}$ for each pair:

$$S_{A/B} = \frac{\text{etch rate of material } A}{\text{etch rate of material } B} \tag{10.18}$$

It is customary to put the material to be removed in the numerator and to express the result as a ratio, e.g. a silicon to SiO_2 selectivity S_{Si/SiO_2} might be $10:1$ in a Si etch process.

Up to about 1975 virtually all etching was done in baths of wet chemicals. This has the virtue of simplicity, although many of the chemicals used present a severe hazard to the operator. Very high selectivities are typical of wet etching, often in excess of $100:1$, both to other layers and to photoresist. Resist failure in wet etching is normally associated with loss of adhesion rather than complete erosion. However, except in a few specific instances where certain crystal planes are attacked preferentially, wet etching is isotropic, i.e. it attacks equally in all directions. As soon as some vertical etching has taken place the exposed sidewall begins to be attacked, leading to a loss of material beneath the mask. The undercut roughly equals the thickness of the film etched (Figure 10.10a). Film thicknesses are typically one micron (10^{-6} m), so wet etching is adequate while linewidths are several microns: the mask feature size can be adjusted to compensate for the undercut. However, when minimum feature sizes approach 2–3 microns wet etching is no longer adequate.

The profile of the etch can be quantified by a new parameter, the anisotropy A, defined as:

$$A = 1 - \frac{\text{lateral etch rate}}{\text{vertical etch rate}} \qquad (10.19)$$

The anisotropy varies between zero, for totally isotropic etching, to one, for a perfectly vertical wall (Figure 10.10b). Control over anisotropy throughout the range has proved possible by etching in an ionized gas or plasma (Figure 10.11). Although gases are good insulators it is possible to ionize a proportion of the atoms or molecules by the application of sufficient dc or rf voltage, especially at reduced pressures. The positive ions and free electrons then give the medium a usable conductivity, such as is found in fluorescent lights.

The electrons in a plasma are much more mobile than the heavier ions and gain energy more quickly from the applied electric field. They pass on the energy to other particles in collisions, which might excite, ionize or fragment

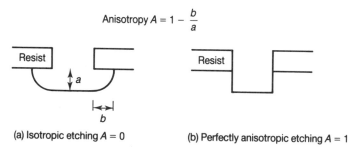

Anisotropy $A = 1 - \dfrac{b}{a}$

(a) Isotropic etching $A = 0$ (b) Perfectly anisotropic etching $A = 1$

Figure 10.10 Extreme etch profiles

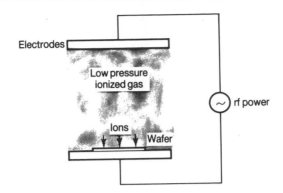

Figure 10.11 Etching in a reactive plasma

them. For example, the following processes occur in a CF_4 plasma:

$$e + CF_4 \longrightarrow CF_3 + F \quad \text{(dissociation)} \tag{10.20}$$

$$e + CF_3 \longrightarrow CF_3{}^+ + 2e \quad \text{(ionization)} \tag{10.21}$$

In this instance a fluorine atom is liberated from the stable CF_4 gas molecule. Fluorine arriving at a silicon surface can etch it, because the compound SiF_4 which forms is volatile. The overall reaction can be summarized:

$$4F + Si \longrightarrow SiF_4$$
$$\text{(surface)} \qquad \text{(gas)} \tag{10.22}$$

Etching by reactive species generated in a plasma is likely to be isotropic. However, surfaces are also subject to ion bombardment. Even a floating surface is negative with respect to the main body of plasma, being charged by fast moving electrons when the plasma is struck. The surface adopts a potential that will balance the arrival of electrons and positive ions thereafter, by repelling electrons and accelerating ions. An important consequence of this is that ions arrive at surfaces almost perpendicularly. In many cases etching will proceed quickly under ion bombardment, but stop or occur slowly on vertical, unstimulated surfaces. This opens the way for anisotropy control, especially if extra accelerating potential can be applied to the wafer.

10.7 Deposition

The methods by which thin films of new material can be added to a wafer will be described in this section. There is some overlap between them – for example, vapour phase epitaxy is a special case of chemical vapour deposition – but the segregation applied corresponds to the classes of equipment used rather than the physical principle employed. A very wide range of organic, inorganic, dielectric, semiconducting and conducting films are deposited routinely, without varying the basic equipment substantially. The topics to be covered are:

1. *Epitaxy*: single crystal material, usually semiconductor.
2. *Chemical vapour deposition*: mostly dielectrics or polycrystalline semiconductor.
3. *Evaporation*: mostly metals.
4. *Sputtering*: metals or insulators.
5. *Spin on*: polymers, organic compounds, and glasses.

The first four techniques all rely on the exclusion of air, usually by operating at reduced pressure in a vacuum system. This is to prevent unwanted oxidation of the deposited film or the source materials. The vacuum requirements can be stringent: consider, for example, an aluminium deposition process occurring at 1 micron/minute. This is equivalent to about 4,000 layers of atoms per minute. For the arrival of oxygen to be 1% of this, an oxygen partial pressure below 10^{-9} atmospheres is necessary.

The total pressure during deposition influences the process through the mean distance between gas particle collisions, known as the mean free path. This is inversely proportional to pressure. If the mean free path is short then the gas flow pattern and gaseous diffusion are important in determining the rate of matter transport. When the mean free path is longer than the dimension of the vacuum chamber, particles travel in straight lines from wall to wall, where they might stick or rebound. The chamber geometry is again an important parameter.

10.7.1 Epitaxy

After going to great lengths to make a pure, single crystal wafer there are still several reasons for wanting to deposit single crystal material on top of it. Such deposition is called epitaxy from the Greek *epi* meaning on and *taxis* meaning ordered. The motivations include:

1. The creation of a highly doped buried layer under lightly doped epitaxy.
2. The formation of multiple junctions.
3. The formation of junctions between different kinds of crystalline semiconductors.

An epitaxial layer of the same material as the wafer substrate is the easiest to achieve, the new layer taking its orientation from the existing material. This is called homoepitaxy. Epitaxial deposition of a different semiconductor – heteroepitaxy – is possible provided the mismatch between the two crystal lattices is small. Heterojunctions between $GaAs_xP_{1-x}$ compounds of different stoichiometry are a good example of this, as illustrated in section 8.8.2 on semiconductor lasers.

The three main techniques for epitaxial deposition are vapour phase epitaxy (VPE), molecular beam epitaxy (MBE) (Figure 10.12), and liquid phase epitaxy (LPE):

1. *Vapour phase epitaxy*: a gaseous compound containing the material to be deposited is brought into contact with the substrate at high temperature (1,000–1,200 °C for silicon). The compound decomposes on the surface,

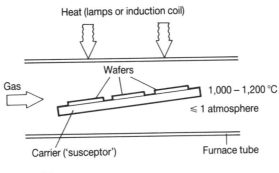

(a) VPE

Note:
Gases decompose on hot surfaces, leaving a solid deposit. A crystalline surface can grow a single crystal film.

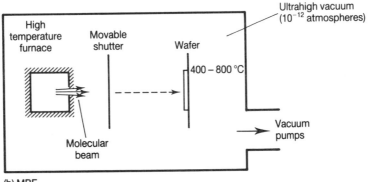

(b) MBE

Note:
The wafer must be clean on an atomic level, requiring ultrahigh vacuum. Epitaxial deposition can occur at much lower temperatures than in VPE.

Figure 10.12 Vapour phase (VPE) and molecular beam epitaxy (MBE)

depositing fresh material, while other reaction products desorb. For example:

$$SiCl_4 \longrightarrow Si + 2Cl_2 \uparrow \qquad\qquad (10.23)$$

The very high temperatures, just below the melting point of silicon (1,410 °C), allow very rapid movement of the newly deposited atoms across the surface. Under these conditions the atoms can try many different sites and are most likely to rest finally in a good crystal lattice, that being a position of minimum energy.

VPE is relatively rapid – around one micron/minute – and is well suited to homoepitaxy. Its main drawback is the very high temperature used, which also drives dopant diffusion very effectively. Although it is possible to deposit doped epitaxial silicon by adding AsH_3 or PH_3 gases it is hard to make sharp multiple junctions because the dopants tend to smear out by diffusion within the growing layer.

2. *Molecular beam epitaxy*: epitaxial growth is possible at 400–800 °C provided the surface is kept scrupulously clean. Deposition takes place by firing a beam of evaporated material at a substrate held at ultrahigh vacuum (UHV: less than 10^{-12} atmospheres pressure). The surface is precleaned under UHV either by heating briefly above 1,000 °C to drive off surface oxides or by chipping away a few surface layers by firing a beam of inert argon ions (sputter etching). The latter is preferred for temperature-sensitive compound semiconductors.

    The film can be doped as it grows by adding a beam of dopants, or by ion implantation. The low temperature means that doping profiles remain almost fixed, enabling complex multilayer structures to be formed. The growth rate is lower, typically 0.1 micron/minute, and the equipment is very expensive: this has restricted MBE to producing special discrete devices at present.

3. *Liquid phase epitaxy*: epitaxial material can be deposited by dipping a seed layer into a melt close to its freezing point, as in the Czochralski process. The method has also been used to deposit epitaxial films, by drawing open floored pots of melt over the wafers. Multiple heterojunctions have been made by LPE, using several melts with differing compositions. Although useful in the laboratory for creating abrupt junctions, the technique has not been widely adopted in mass production.

10.7.2 Chemical vapour deposition (CVD)

The principle of CVD is outlined in Figure 10.13: a molecule decomposes or reacts on a surface. Some of the reaction products are sufficiently volatile to

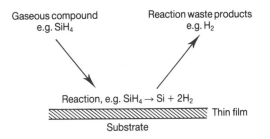

Figure 10.13 Chemical vapour deposition (CVD)

leave the surface and are pumped away, while other products remain on the surface as a growing layer. Films containing silicon can be obtained by decomposing silane (SiH_4) or a chlorosilane (using SiH_2Cl_2 or $SiCl_4$). Fully or partially oxidized silicon can be deposited by adding an oxidizer (O_2 or N_2O) to the gas mixture; silicon nitride requires ammonia (NH_3) or another gas which can form atomic nitrogen.

Metal films can also be deposited, provided a suitable metal-bearing gas or vapour exists. The reactions of most interest to semiconductor device fabrication at present are two processes for tungsten deposition:

$$2WF_6 + 3Si \longrightarrow 2W + 3SiF_4 \tag{10.24}$$

$$WF_6 + 3H_2 \longrightarrow W + 6HF \tag{10.25}$$

In the first reaction tungsten hexafluoride is reduced by silicon, depositing tungsten. This is a selective process, occurring only where silicon is exposed. The hydrogen reduction process can also be selective, proceeding only where metal has already been laid. Taken together, the reactions offer the possibility of saving one photolithographic sequence.

The deposition rates are fairly slow in most CVD processes, ranging from 30–300 nm/minute. Systems which can process several wafers simultaneously are preferred in these circumstances. A great advantage of CVD is the conformal deposition frequently obtained even over steep steps (Figure 10.14a), which is in contrast to the variation in thickness obtained by evaporating or sputtering a thin film.

There are many variants on the CVD technique, distinguished by the pressure range and the energy source driving the chemical reaction:

1. ACVD, APCVD: atmospheric pressure CVD. The reactants are transported in a stream of inert carrier gas and are decomposed thermally.

2. LPCVD: low pressure CVD, usually $<10^{-3}$ atmospheres. A carrier gas is not necessary but the vacuum pump must be able to handle all the reactants and gaseous products. Uniformity across a batch of wafers is easier to achieve at reduced pressure. Reactions are thermally driven.

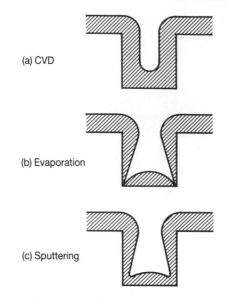

(a) CVD

(b) Evaporation

(c) Sputtering

Figure 10.14 Typical step coverage

3. PECVD: plasma enhanced CVD. This falls within LPCVD but uses a plasma to increase the deposition rate at a lower temperature, usually less than 400 °C.

4. LCVD: laser CVD. The reaction is triggered by a combination of photon absorption and laser heating.

10.7.3 Evaporation

A typical evaporation apparatus is shown in Figure 10.15. It consists of a water-cooled hearth containing a charge of source material, heated in this case by an electron beam directed by a magnetic field. High vacuum is used to prevent oxidation of the source or the deposited film, and to allow evaporating material to move freely to the cooler wafer and chamber surfaces, where it sticks. Wafers are held in a planetary motion drive to improve both uniformity across the wafers and the step coverage. If a wafer is held fixed material will arrive from only one angle, causing breaks in the film where a step casts a shadow. Rotation improves this but cannot cure it because the growing film itself casts a shadow. This leads to a profile such as that shown in Figure 10.14b. Heating the wafers during deposition also helps, by promoting surface movement in the film.

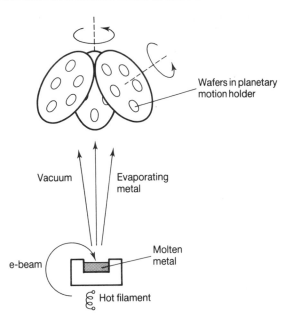

Figure 10.15 Evaporation system schematic

Low melting point metals such as aluminium are the most suitable for evaporation. Refractory metals and insulators can be evaporated, but with more difficulty. Alloys such as Al/1% Si can also be deposited, the problem here being consistency of composition during the life of the charge, because aluminium evaporates more readily. In its favour, the technique gives rapid deposition rates of up to several microns per minute.

10.7.4 Sputtering

The heart of the sputtering process is illustrated in Figure 10.16. An inert gas plasma is struck between two electrodes; wafers to be coated are placed on one, and a target of source material is bonded to the other. Positive ions from the plasma strike the target with sufficient energy to chip away atoms or clusters of atoms. The pressure in the plasma is typically 10^{-6} atmospheres, so the material released makes few or no collisions before striking a surface and sticking. Deposition is therefore line of sight, as in evaporation, but the larger source area can give an improvement in step coverage (Figure 10.14c).

Virtually any material can be sputtered, including high melting point metals, insulators and semiconductors. Compounds and alloys can be sputtered, although the different sputter rates of the components make compositional

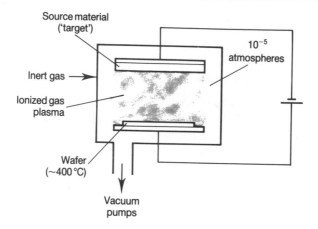

Note:
Positive ions from the plasma chip atoms off the target, which then stick to other surfaces.

Figure 10.16 Sputter deposition

control of the deposited film more difficult. Sputtering is currently the dominant method for depositing metal films. Variants include the following:

1. *Dc or rf sputtering*: either can be used if the target is conducting, while dielectric sputtering requires rf excitation.

2. *Magnetron sputtering*: permanent magnets or field coils are commonly placed behind the target to intensify the plasma locally and increase the sputter deposition rate.

3. *Reactive sputtering*: a reactive gas can be used in the plasma to contribute material to the film. For example titanium nitride can be deposited by sputtering titanium in nitrogen.

4. *Bias sputtering*: an independent bias is applied to the wafers for one of two reasons:
   (a) to sputter away some of the deposited film – this can redistribute materials from the floor of grooves to the sidewalls, and give control over step coverage, at the cost of reducing the deposition rate;
   (b) to bombard the growing metal film with energetic ions, and cause partial melting or enhanced surface mobility. Surface tension causes material to reduce curvature, and can lead to filling of small holes. Such planarization makes subsequent deposition and photolithography much easier.

10.7.5 Spin on

Some materials, notably photoresist and organic dielectrics such as polyimide, can be dispensed in liquid form on to a wafer. If the wafer is then spun at a few thousand rpm for 10–20 seconds, excess liquid whirls off leaving a uniform coating (Figure 10.2). The thickness of the coating will depend on the spin speed and the liquid's viscosity. Solvents are then driven off in a temperature cycle, leaving a thin solid film. The technique is relatively simple and rapid, and usually results in a high degree of planarization or smoothing out of the surface.

10.8 Circuit integration

The power of semiconductor devices lies not so much in the characteristics of any single circuit element as in the possibility of combining many elements in a complete circuit on a single piece of semiconductor. Each circuit is constructed on a chip or a die ranging in size from 0.5 mm × 0.5 mm or less up to 10 mm × 10 mm, so that several hundred die can be fabricated on a single wafer (Figure 10.17). Nearly all processing steps treat a whole wafer or even a batch of wafers simultaneously, so that mass production can take place on a truly grand scale. A semiconductor fabrication line can produce a given range of devices – diodes, resistors, capacitors and active transistors – side by side on a wafer surface. The integrated circuit designer combines these elements into a circuit to perform a required electronic function, such as amplification, digital data storage or manipulation, or filtering. The design will often call up subsystems designed previously, using computer-aided design tools (CAD). The patterns of each layer making up the circuit are transferred on to photolithographic mask plates, usually by electron-beam writing (lithography). The power of integrated circuit fabrication is that the manufacture is nearly *independent of the pattern*. The circuit can be changed by using different mask plates, while the whole sequence of oxidations, doping, layer creation and etching remains fixed.

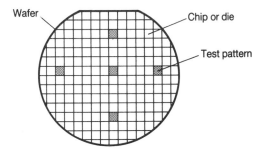

Figure 10.17 Each wafer consists of a large number of identical patterns

There are several motivations for packing more and more circuitry on a single die:

1. Equipment becomes less bulky.
2. Reliability is improved by reducing the number of solder or plug and socket interconnections.
3. Assembly and packaging costs are minimized.
4. Circuit performance improves: in particular, higher digital clock speeds are possible if the interconnection path lengths are reduced and the individual devices are made smaller.

There are four ways of making devices smaller:

1. Make the individual devices smaller. This has called for better lithography tools capable of producing smaller linewidths with more accurate layer to layer registration. Five micron minimum feature sizes were common around 1975. This became two microns during 1980–85. The highest packing densities currently in use (1990) call for 0.8 micron features. This trend is expected to continue down to about 0.1 micron features, to a point at which the active device characteristics change for the worse.

2. Make the die bigger. The limit here is the trade off between die size and the fractional yield of working devices. All stages of manufacture can add a few defects to the wafer, usually as particles embedded on the surface. The larger a die is, the more likely it is to contain a killing defect. This can be roughly expressed as a Poisson probability: if the die area is A m^2 and the defect density is D m^{-2}, the yield fraction Y of good die is given by:

$$Y = \exp{-DA} \qquad (10.26)$$

By this reckoning, doubling the die area reduces the yield to 37% of its former value, which almost triples the price of producing a good circuit. (This turns out to be too conservative in practice, because of a tendency for defects to cluster. In this case a plausible theory retarded technological advance.)

3. Waste less space. We have referred already to the move away from wet isotropic etching towards vertical profiles which take up less room. Another concern is the space used to isolate one device from others on the same die. There is also a tendency to increase the depth of the device by adding three or more layers of interconnection patterns, rather than spreading out laterally.

4. Use less devices to do the same job. It used to take six devices to store one

binary bit of data: two are now sufficient. Space-saving circuit tricks fill whole journals.

Two other trends should be noted. First, wafer size keeps increasing. Wafers of diameter 25 mm and 50 mm were used in the 1960s, while 100–150 mm and even 200 mm silicon wafers are in use now. It costs about the same to process a wafer regardless of its size, so there is a strong economic pressure towards larger wafers. The limit here is not so much technological as economic: beyond a certain size each wafer becomes too valuable to lose by bad processing or breakage, especially when it is nearly complete.

Second, there is a move away from silicon toward GaAs and other semiconductors. The reason becomes clear when the electron mobilities are compared: much faster devices can be made in GaAs, using the same feature size. At present silicon holds around 90% of the market, and it will continue to be the dominant material for some decades unless high quality GaAs substrates become much cheaper and more readily available.

10.9 Passive components

10.9.1 Isolation

There must be a means of electrically isolating individual circuit elements if a complete circuit is to be fabricated on a single die. The semiconductor surface is usually passivated by growing a thick silicon dioxide layer, which also acts as an insulator. This layer separates conducting tracks from the devices, except where windows are opened to make contacts. The insulator must be thick enough to stop a conducting channel being induced in the semiconductor by the potential applied to the upper conductor. (This will be elaborated in Chapter 14.)

Isolation within the substrate usually requires the formation of a reverse biased *pn* junction around the device. Figure (10.18) illustrates this: the *p*-type substrate is connected to the most negative circuit potential, and other *p*-type regions are all surrounded by *n*-type regions at more positive potentials. Junction isolation works well provided the substrate leakage currents are small compared to the normal circuit currents.

10.9.2 Resistors

Resistors can be made in the doped substrate, as in Figure 10.19 or by depositing resistive material – usually nichrome metal alloy or doped polycrystalline silicon (polysilicon or just poly). The substrate resistor is also useful as a cross-over position, without using two interconnection layers. Values between 50 ohms and 10 kilohms can be made routinely on a chip: other values require

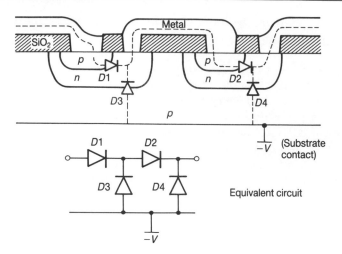

Note:
Diodes $D3$ and $D4$ are reverse biased if $-V$ is the most negative circuit potential.

Figure 10.18 Junction isolation

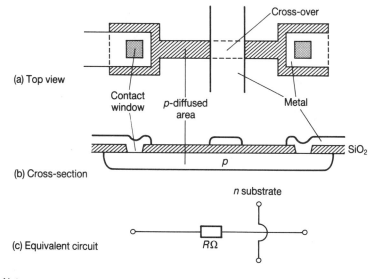

Note:
In (a) the nesting of the contact window within the metal strip allows for possible misalignment.

Figure 10.19 Diffused resistor and cross-over

an external component and should be avoided if possible. The absolute resistance is hard to control to better than ±5%, but the ratio of resistances on a die can have a much tighter tolerance. This has a strong influence on circuit design. A precise resistor value can be obtained by laser trimming a metal stripe: the laser nibbles metal by evaporation until the resistance rises to the required value.

The resistance of thin films or doped surfaces to lateral current flow is usually quoted as a *sheet resistance* R_s in ohms per square. Consider for example the stripe in Figure 10.20 of width L, length $n \times L$, thickness t and uniform resistivity ρ_e. n is a number, not necessarily an integer. Its resistance, R, is given by

$$R = \frac{nL\rho_e}{Lt} = \frac{n\rho_e}{t} = nR_s \; \Omega \tag{10.27}$$

R_s takes this definition for a layer of uniform resistivity, but must be defined by:

$$R_s = \frac{\text{stripe resistance} \times \text{stripe width}}{\text{stripe length}} \; \Omega/\text{square} \tag{10.28}$$

when the resistivity varies with depth.

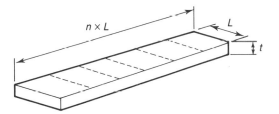

Figure 10.20 Resistance and sheet resistance

EXAMPLE

■ Determine the resistance of stripes of dimension (length × width) 100×10 microns; 200×20 microns; 100×6 microns, if the sheet resistance is 50 ohms/square.

☐ Rearranging equation (10.28):

$$\text{stripe resistance} = R_s \times \frac{\text{length}}{\text{width}}$$

The length : width ratio for the 100×10 and 200×20 stripes is ten for both, i.e. it can be divided into ten square segments. The resistance is then

$10R_s = 500$ ohms. The third stripe is $100/6 = 16.67$ squares long, so its resistance is:

$$50 \frac{\text{ohms}}{\text{square}} \times 16.67 \text{ squares} = 833 \text{ ohms}$$

10.9.3 Capacitors

There are two ways of making a capacitor on a chip:

1. Reverse bias *pn* junction: this gives a small-signal ac capacitance dependent on the size of the dc applied voltage $|V_A|$ across the junction. The capacitance, C, varies as V_A^{-n}, where $n = 0.5$ for an abrupt junction and 0.33 for a linear junction.

2. A conductor–dielectric–(semi-)conductor capacitor: typically this would take the form shown in Figure 10.21, with interconnection metal forming the top plate, silicon dioxide the dielectric and a heavily doped semiconductor layer acting as a counter electrode. Much smaller trench capacitors can be made by etching a deep, narrow trench, oxidizing the walls and depositing a counter electrode which fills the groove. This metal–oxide–semiconductor (MOS) capacitor also forms the heart of a large class of active devices, to be discussed later.

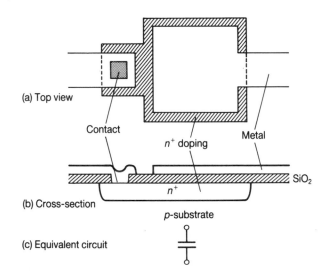

(a) Top view

Contact

n^+ doping

Metal

(b) Cross-section

n^+

p-substrate

SiO_2

(c) Equivalent circuit

Figure 10.21 Integrated capacitor

Whichever system is used, the range of values accessible on the chip is about 1–50 pF. Larger values are possible but begin to take up too much space. This increases both the chance of finding a defect in the capacitor and the cost of the capacitor. If a larger value is required it must be connected externally.

Inductors are virtually never constructed within integrated circuits because plane geometry dictates that only very low values of inductance can be made in the form of a flat spiral. An external component could be used but more often the circuit is designed from active semiconductor devices, resistors and capacitors only.

10.10 Example of circuit integration

A manufacturing sequence can now be devised for a circuit such as the one in Figure 10.22. This is a peak detection circuit, the capacitor C charging through the diode $D1$ to the highest voltage applied across the input side (less the turn on voltage of $D1$). The resistors $R1$ and $R2$ discharge the capacitor in a time of order $(R1 + R2)C$ seconds so that it can detect the next peak. They also divide the capacitor potential by the fraction $R2/(R1 + R2)$. The diode $D2$ stops the input terminal going negative by more than one diode drop and so protects $D1$ against breakdown.

The simplest construction sequence uses a p-type silicon wafer, which will be the ground plane, and uses diffusion doping. One possible layout is shown in Figure 10.23:

1.   Grow 1 micron silicon dioxide in a wet furnace process.

2.   Photolithography: define regions for n-type diffusions.

3.   Etch n-type diffusion windows in SiO_2; strip resist.

4. n-type diffusion, predeposition and drive in: this creates the n-type part of the diodes, and the resistor isolation. The drive in can be performed in an

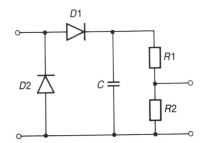

Figure 10.22 Example circuit for integration

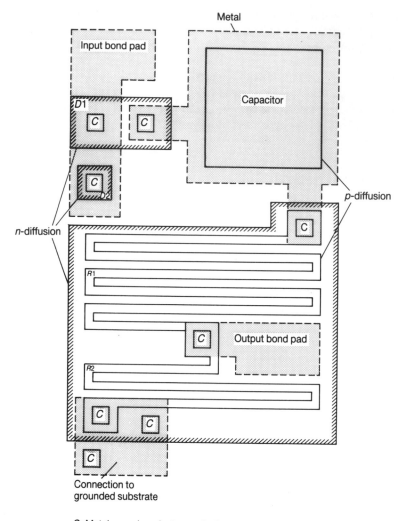

C: Metal – semiconductor contact

Figure 10.23 Integrated layout of the example circuit

oxidizing atmosphere to regrow some silicon dioxide. It will not be as thick as the initial oxide, which will continue to grow, so the n-diffusion areas are still visible. This is important to allow the next process to be aligned properly.

5. Photolithography: define regions for p-type diffusion – the p-side of diode $D1$, the capacitor region, and the resistor path.

6. Etch p-diffusion windows in the SiO_2; strip resist.

7. p-type diffusion: this must be shorter than the n-type diffusion to avoid swamping it.

8. Grow thin SiO_2 layer in a dry furnace process, to passivate the wafer. This covers the resistor and forms the dielectric of the capacitor.

9. Photolithography: define contact windows to the p-side of $D1$, the n-side of $D2$ and the three contact points on the resistor meander.

10. Etch contact windows in SiO_2; strip resist.

11. Deposit Al metal layer over the wafer by evaporation or sputtering.

12. Photolithography: define the areas of metal required to form the circuit interconnections and the top electrode of the capacitor.

13. Etch unwanted areas of metal; strip resist.

14. Deposit silicon nitride coating layer over the wafer by CVD. This seals the surface against contamination and provides a degree of mechanical protection for the wafer.

15. Photolithography: define bond pad windows.

16. Etch silicon nitride in the bond pad areas, where contact will be made to the external circuitry.

17. Thin the wafer by grinding or etching the rear surface.

18. Test each die on the wafer by lowering small metal probes on to the bond pads and exercising the circuit electrically. Mark each failure with a dot of magnetic ink. Testing is done at this stage because the packaging will typically cost the same as the die.

19. Scribe the wafer between each die and saw it to separate the die. Magnetically remove the bad die.

20. Bond the good die to individual headers, which consist of a metal base with wire feedthroughs. The ground connection is made at this point.

21. Make wire bonds between the bond pads on the die and the wire connections. A fine gold or aluminium wire is welded ultrasonically first on

the bond pad, then on the external connection point. A drop of silicone rubber or epoxy is put over the die immediately, to protect the delicate bond wires.

22.   The top is sealed on the header, completing the construction.

Many different packaging styles are available, depending on the number of interconnection wires and the environment the integrated circuit must work in. The commonest is the dual in line package, featuring two rows of legs on a pitch of 0.1 inches, 0.3 inches apart, which is suitable for a maximum of 60 interconnections. This can be increased by siting connections around the periphery of a square package, with a smaller space between legs, or by using a pin-grid array of connection legs from the whole base area. The trend is to put more and more of a system on a single die, which usually increases the number of inputs and outputs required. Lead lengths are shrinking as digital clock speeds increase, because of the time lags due to the inductance and capacitance of the leads. A significant problem in many packages is heat dissipation: faster switching often means a higher power consumption, which must not be accompanied by an excessive die temperature. Simple packages can dissipate up to a few watts but more expensive versions can extend this to tens of watts.

Summary

Main symbols introduced

Symbol	Unit	Description
A		anisotropy of etch
A	m	first oxidation coefficient
B	$m^2\,s^{-1}$	second oxidation coefficient
d_{ox}	m	oxide thickness
D	m^{-2}	defect density
D	$m^{-2}\,s^{-1}$	dopant diffusion coefficient
D_0	$m^{-2}\,s^{-1}$	pre-exponential diffusion constant
E_a	J	activation energy for diffusion
Q	m^{-2}	dopant dose
R	m	ion implantation range
R_s	$\Omega\,square^{-1}$	sheet resistance
$S_{A/B}$		etch selectivity of A over B
σ_R	m	standard deviation of ion implantation range
τ_0	s	third oxidation coefficient

Main equations

$$D = D_0 \exp - E_a/kT \text{ m}^2\text{s}^{-1} \quad \text{(dopant diffusion)} \tag{10.1}$$

$$\frac{\partial N}{\partial t} = D\frac{\partial^2 N}{\partial x^2} \text{ m}^{-3}\text{s}^{-1} \quad \text{(Fick's second law)} \tag{10.5}$$

$$\text{erf}(z) = (2/\sqrt{\pi}) \int_0^z \exp(-y^2)\text{d}y \quad \text{(error function)} \tag{10.7}$$

$$\text{erfc}(z) = 1 - \text{erf}(z) \quad \text{(complementary error function)} \tag{10.8}$$

$$d_{ox} \simeq (B/A)(t + \tau_0) \text{ m} \quad \text{(thin oxide)} \tag{10.15}$$

$$d_{ox} \simeq (Bt)^{1/2} \text{ m} \quad \text{(thick oxide)} \tag{10.16}$$

$$R_s = \frac{\text{stripe resistance} \times \text{stripe width}}{\text{stripe length}} \Omega\,\text{square}^{-1} \tag{10.28}$$

Exercises

1. The following statements are either true or false:
    (a) Impurity levels in virgin semiconductor wafers can exceed device doping levels.
    (b) Arsenic is readily lost from GaAs when it is heated.
    (c) The exposed parts of positive resist are removed by the developer.
    (d) Isotropic etching will tend to compensate for light creeping under mask features in negative resist photolithography. (Draw some pictures.)
    (e) Diffusion of dopants will eventually destroy all *pn* junction devices.
    (f) Drive in diffusion leaves the surface doped to the solid solubility limit.
    (g) The total amount of dopant in the semiconductor is constant during a predeposition diffusion.
    (h) Ion implantation allows independent control of total dopant dose and junction depth.
    (i) Diffusion doping spreads sideways under the mask more than ion implantation doping.
    (j) The thermal oxidation rate of thick SiO_2 films is limited by the diffusion of oxygen through the film.
    (k) Wet etching proceeds faster vertically than under the mask.
    (l) The anisotropy can exceed one in plasma etching.
    (*m*) Thin films of refractory metals can be deposited by sputtering more easily than by evaporation.
    (n) Epitaxial deposition produces polycrystalline films.
    (o) Thin film deposition is performed under vacuum to keep dust out.
    (p) Heteroepitaxy requires crystal lattices of similar size.
    (q) An even coating over steps is easier to achieve by chemical vapour deposition than by sputtering.

(r)   Magnetron sputtering uses microwaves to make a plasma.

(s)   Smaller dies will tend to have a larger yield.

(t)   Anisotropic etching of films over vertical steps requires a high selectivity because of the material on the side of the step.

(u)   Chips can be made more cheaply using larger wafers.

(v)   Junction isolation is a means of joining devices together by the substrate.

(w)   Resistance ratios have a tighter tolerance than absolute values in an integrated circuit.

(x)   Doubling the film thickness doubles the sheet resistance.

(y)   It is better to test the completed circuit after it has been packaged and not before.

2.   Sketch a top view and cross-sectional diagram sequence illustrating a negative photoresist process to open a diffusion window in a silicon dioxide layer.

3.   A dry oxidation process with an initial oxide thickness of 20 nm takes ten minutes to grow a 22.5 nm oxide and 24 hours to grow a 268 nm oxide. How long will it take to grow a 0.1 micron oxide by this process? Suggest two ways of shortening this time.

4.   The diffusion length for a 30 minute process at 1,050 °C is 0.48 microns, and the diffusion activation energy is 3.1 eV. Find the diffusion time which gives the same doping profile at 1,000 °C.

5.   A predeposition diffusion is performed using phosphorus, with a surface concentration of 10^{27} m^{-3}. The substrate is doped p-type with a uniform acceptor ion density of 10^{21} m^{-3}. Find the depth of the pn junction below the surface after a 1,000 second diffusion if the diffusion coefficient is 2.10^{-17} m^2 s^{-1}.

6.   A film requiring etching is not completely uniform so that the maximum and minimum thicknesses for etching are $A + a$ and $A - a$ respectively ($A > a$). Likewise the etch rate is not uniform over the wafers, but varies between $V + v$ and $V - v$. The maximum amount of underlying material that can be removed is L. Show that the nominal selectivity must be at least:

$$\frac{2V}{L} \frac{(Av + aV)}{V^2 - v^2}$$

Evaluate this for a film 1 micron thick ±3% with an etching rate of 0.2 microns/minute ±2%, if a maximum of 0.02 microns can be etched from the substrate.

7.   Devise a process sequence for creating highly doped local regions underneath a layer of crystalline semiconductor, using a combination of diffusion doping and epitaxial deposition. (This is the buried layer process.)

8.   Suggest a deposition method for each of the following films, bearing in mind the constraints given and the need to maximize throughput:

(a)   Al : Cu : Si = 94.5% : 4% : 1.5%, 1 micron thick;

   (b)   silicon nitride, 0.2 microns thick, at a maximum temperature of 300 °C;
   (c)   Al 1 micron thick, good step coverage not required;
   (d) $Al_xGa_{1-x}As$ and $Al_yGa_{1-y}As$ in alternating layers 0.05 microns thick;
   (e)   polycrystalline silicon 0.7 microns thick with good step coverage, at a
          maximum temperature of 800 °C.

9.   An existing manufacturing process has an overall yield of 61% and consists of 12
      stages:
      (a)   find the yield per stage, assuming it is the same throughout manufacture;
      (b)   find the overall yield if three more identical stages are added;
      (c)   calculate the percentage reduction required in defect density if the 15-stage
             process is to achieve the same yield as the 12-stage process.

10.   Devise a process sequence for fabricating the capacitor structure in Figure 10.21.
      Illustrate your answer with cross-sectional diagrams.

11.   Sketch a top view of the doping areas and metal tracks required to create the diode
      logic circuit shown in Figure 10.24 in an n-type silicon wafer. Describe the process
      sequence up to and including the metal pattern definition.

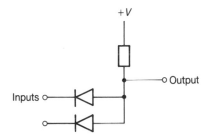

Figure 10.24 Diode logic circuit

11 Bipolar junction transistors – dc

11.1 Description

The 1940s witnessed the birth of two world-changing inventions: nuclear explosive devices and bipolar junction transistors. The rest of this century may be construed as a race between the destructive capacity of the one and the growth in planet-wide communication spawned by the other. The bipolar junction transistor (or BJT) took over two tasks from the bulky, fragile, high voltage valve:

1. *Amplification*: it can magnify small electrical oscillations.
2. *Switching*: it can be turned from an on conducting state to an off blocking condition.

In this chapter we will give a description of the BJT and an explanation of its primary electrical characteristics. The next chapter will deal with its responses to rapid oscillations and switching transients. The BJT consists of two *pn* junction diodes back to back, close together in the same semiconductor crystal (Figure 11.1). Both *n–p–n* and *p–n–p* configurations are possible. Ohmic metal contacts

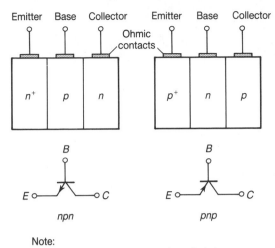

Note:
The arrow follows the base – emitter diode type.

Figure 11.1 Bipolar junction transistors

are made to each part of the sandwich. The outer layers are called the emitter and collector, while the central region is the base. The emitter and collector are not normally interchangeable because the emitter doping is usually much heavier. Figure 11.1 also shows the circuit symbols: the arrow corresponds to the base–emitter diode polarity.

11.2 Fabrication

BJTs can be made either as discrete devices or in planar integrated form. In a separate device the substrate can be used for one connection, typically the collector (Figure 11.2a). All three contacts appear on the top surface in the integrated version and the substrate must make a reverse biased diode with the collector (Figure 11.2b). The emitter–base diode is closer to the surface than the base–collector junction because it is easier to make the heavier emitter doping at the top. Current can flow readily between the emitter contact and the first junction but must take a more tortuous path between the other contacts and the active parts of the transistor directly underneath the emitter. The series resistances are called the base spreading resistance and the collector internal resistance.

The buried layer process outlined below was introduced to minimize the collector internal resistance and to avoid the technical difficulties encountered in forming the collector, base and emitter by three sequential diffusions. It relies

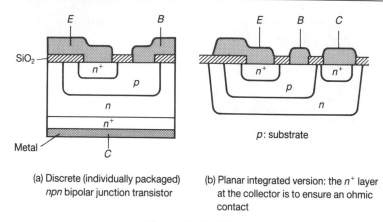

(a) Discrete (individually packaged)
npn bipolar junction transistor

(b) Planar integrated version: the n^+ layer at the collector is to ensure an ohmic contact

Figure 11.2 BJTs

on epitaxial deposition to create a highly doped region underneath the transistor. A typical *npn* fabrication sequence is outlined in Figure 11.3:

1. Oxidize a *p*-type substrate.

2. Photolithography and etching to open windows where transistors are required.

3. Diffuse n^+ (Figure 11.3a).

4. Oxidize: the growth rate is faster in the exposed, doped areas, so more silicon is converted to SiO_2 in the windows (Figure 11.3b).

5. Strip away all silicon dioxide. A small step remains which is used to align subsequent photolithographic steps. For clarity, the step is not shown in subsequent diagrams.

6. Deposit lightly doped *n*-type silicon epitaxially. The doping level is that required for the collector.

7. Oxidize the surface of the epitaxial layer (Figure 11.3c).

8. Photolithography and etching to define a ring-shaped window around the transistor.

9. Diffuse *p*-type through the thickness of the epitaxial layer, re-forming the oxide by doing the drive in diffusion in an oxidizing atmosphere (Figure 11.3d). This creates an island of *n*-type crystalline semiconductor, surrounded, laterally and below by *p*-type material for junction isolation.

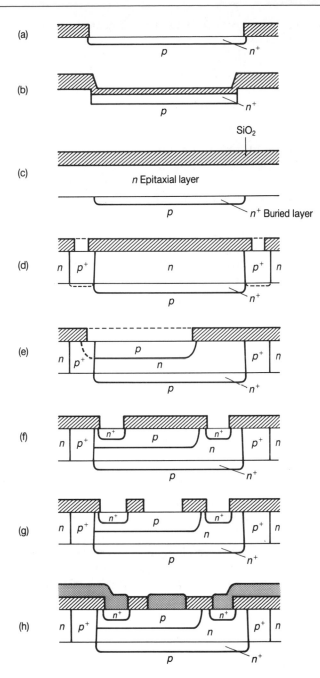

(a)

(b)

SiO$_2$

(c)

n Epitaxial layer

n^+ Buried layer

(d)

(e)

(f)

(g)

(h)

Note:
The various oxide layers are shown with constant thickness where the detail is uneccessary.

Figure 11.3 Buried layer process for manufacturing integrated BJTs

An *npn* transistor can now be fabricated in this region by two more diffusions.

10.   Photolithography and etching to open a window for the base diffusion.

11.   Diffuse base *p*-type, by predeposition and drive in with oxidation (Figure 11.3e).

12.   Photolithography and etching: open diffusion windows for the collector contact area and the emitter.

13.   Diffuse n^+ emitter and collector contact (Figure 11.3f).

14.   Photolithography and etching to open base contact area. The emitter and collector contacts are already open and require only a brief etch to remove a thin oxide formed in (13). This can be done without the need for further masking, so these contacts are self-aligned with the n^+ diffusion areas (Figure 11.3g).

15.   Deposit metal and heat treat to make ohmic contacts to emitter, base and collector.

16.   Photolithography and etching to define the interconnection pattern. The BJT is complete, requiring only packaging (Figure 11.3h).

A feature of this process is the possibility of making *pnp* transistors using the same diffusions. The cross-sectional and plan diagram of a lateral *pnp* BJT is given in Figure 11.4. The collector forms a partial ring around the emitter, both formed from the base diffusion of the *npn* transistor. The base doping is that of the deposited epitaxial layer and the buried layer reduces the base spreading resistance. The separation of base and emitter is important, as will become clear later: charges injected from the emitter into the base must be captured by the base–collector diode in order to obtain transistor action. The lateral *pnp* transistor has relatively poor characteristics because the emitter is not heavily doped. Nevertheless, the ability to make both *npn* and *pnp* transistors in one sequence offers a lot of flexibility to the circuit designer.

11.3 Transistor action

11.3.1 Bias options

The emitter–base and base–collector diodes offer four biasing possibilities, summarized in Table 11.1. The cut off and saturation states follow readily from

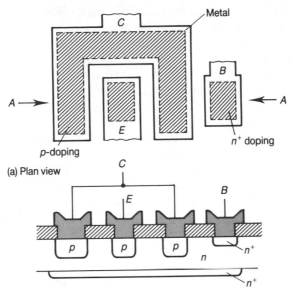

(a) Plan view

(b) Cross-section along A – A

Figure 11.4 Lateral *pnp* transistor, made using the same doping processes as the buried layer *npn* BJT

Table 11.1 Bipolar junction transistor biasing

Bias combination		Condition	Emitter–Collector current flow
E–B diode	B–C diode		
Reverse	Reverse	Off or cut off	Leakage only
Forward	Reverse	Transistor action	Controlled by base current–high gain
Reverse	Forward	Inverse transistor action	Controlled by base current–low gain
Forward	Forward	On or saturation	Uncontrolled high current

considering the device as two diodes back to back and are the basis of digital bipolar circuits. We will concentrate on the normal bias condition which results in transistor action:

base-emitter forward biased
base-collector reverse biased

In this state the current flow between emitter and collector is controlled by a much smaller current flowing at the base contact and this is called transistor action. (Read that sentence again – it is important.) This is not the only possible way of looking at the BJT but it is the foundation of the most common circuit applications. First we will go through a mathematical analysis of transistor action, considering the electron and hole flows inside the BJT, which will attempt to justify the definition above. Then in section 11.4 a more physical description will be provided.

Figure 11.5 defines the dc voltage and current symbols and conventions for a BJT. Note that the definitions apply to both a *pnp* and an *npn* transistor, without changing any directions or polarities. The voltage designations are used in many texts but the current directions do vary from book to book: we are taking positive charge flow into the device as positive current.

11.3.2 Electron and hole flows

The electron and hole flows in an *npn* BJT biased for normal operation are illustrated in Figure 11.6. Electron flows are given as solid arrows and hole flows as dotted arrows. The size of the arrow shows the relative size of the particle flow. (A supernatural revelation is not essential to derive such a diagram, provided we bear in mind what is already known about *pn* junctions.) First we will describe each flow, then Kirchoff's current law will be applied to the emitter, base, collector and whole device in turn to derive relations among the currents.

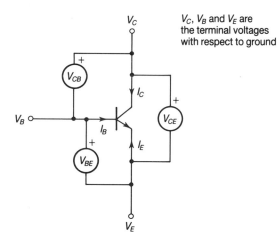

Note:
The '+' signs indicate the voltmeter positive terminals. V_{CB}, for example, could be positive or negative.

Figure 11.5 BJT voltage and current conventions for both *npn* and *pnp* devices

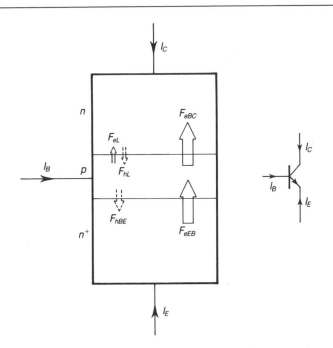

Figure 11.6 Charge flows in a normally biased *npn* transistor

1. F_{eEB}: electron flow from emitter to base. The emitter–base diode is forward biased, so the barrier to electron diffusion from the n^+ doped emitter to the p-type base is reduced, producing a large flow of electrons.

2. F_{hBE}: hole flow from base to emitter. Forward bias also encourages this flow, but it is much weaker than F_{eEB} if the base doping is small compared to the emitter doping. For example if the emitter doping is one hundred times the base doping the forward bias barrier reduction will allow about one hundred electrons to be injected from emitter to base for every hole crossing in the opposite direction.

3. F_{eBC}: electron flow from base to collector, *originating in the emitter*. This is the key to how a transistor works. The band diagram in Figure 11.7 for the normally biased *npn* transistor shows that there is a favourable energy gradient for electrons in the base to be swept into the collector. Electrons in the base of an *npn* transistor are minority carriers and can expect to recombine after being injected from the emitter. However, *if* the base is narrow enough *and* the collector–base junction is reverse biased, *most* of the injected electrons cross into the collector. This flow far outweighs the next two, which are the leakage fluxes of the reverse biased collector–base diode.

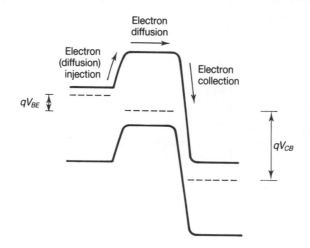

Figure 11.7 BJT band diagram: *npn* transistor in normal bias

4. F_{eL}: minority electrons normally resident in the *p*-type base flow into the collector.

5. F_{hL}: minority holes in the *n*-type collector flow into the base.

Forward biasing the emitter–base junction greatly increases the number of minority carriers in the base. These are received by the collector region in addition to the usual tiny flow of minority carriers. If the base is much less heavily doped than the emitter, then $F_{hBE} \ll F_{eEB}$, and the *dominant* charge flow is from the emitter, through the base, and into the collector.

11.3.3 Kirchoff's current law

Particle flows are all very fine but how do they do anything useful in a circuit, like amplification? To understand this, we must derive some relations among the flows, using Kirchoff's current law: *the currents flowing into a node (or a region) must sum to zero*. Our work is complicated by the fact that we have to consider flows of both positive and negative charges. Positive charge flowing into a region is positive current, so negative charge flowing out of a region is also positive current. The charge particle flows defined above can be converted to currents by multiplying by $q(q = +1.6 \times 10^{-19}$ C), and then Kirchoff's law can be applied:

$$I_E + I_B + I_C = 0 \quad \text{(whole device)} \tag{11.1}$$

$$I_E + qF_{hBE} + qF_{eEB} = 0 \quad \text{(emitter)} \tag{11.2}$$

$$I_B + qF_{hL} + qF_{eL} + qF_{eBC} - qF_{eEB} - qF_{hBE} = 0 \quad \text{(base)} \tag{11.3}$$

$$I_C - qF_{hL} - qF_{eL} - qF_{eBC} = 0 \quad \text{(collector)} \tag{11.4}$$

EXERCISE

Look at Figure 11.6 and write down the above equations. Then looking only at the equations, sketch the figure. This is worth doing until you score 100% – it always comes up in examinations.

11.3.4 Internal BJT parameters

Some of the quantities in equations (11.1)–(11.4) can be lumped together by defining other parameters. The collector–base leakage fluxes qF_{hL} and qF_{eL} only appear together and can be given the symbol I_{CB0}, for the reverse saturation current of the collector–base diode:

$$I_{CB0} = qF_{hL} + qF_{eL} \tag{11.5}$$

The *emitter injection efficiency* γ is defined by

$$\gamma = \frac{F_{eEB}}{F_{eEB} + F_{hBE}} = \frac{-qF_{eEB}}{I_E} \tag{11.6}$$

The hole flow F_{hBE} is wasted as far as the transistor is concerned and only flows in one diode. The electron flow F_{eEB} is the useful injection flow, so γ is the ratio of this flow to the total flow.

The *base transport factor*, b, is the fraction of injected electrons which cross into the collector:

$$b = \frac{F_{eBC}}{F_{eEB}} \tag{11.7}$$

The remaining fraction $(1 - b)$ recombines in the base. The four current equations (11.1)–(11.4) become:

$$I_E + I_B + I_C = 0 \tag{11.8}$$

$$I_E + qF_{eEB}/\gamma = 0 \tag{11.9}$$

$$I_B + I_{CB0} + qF_{eEB}\left(b - \frac{1}{\gamma}\right) = 0 \tag{11.10}$$

$$I_C - I_{CB0} - qF_{eEB}b = 0 \tag{11.11}$$

Eliminating F_{eEB} between (11.9) and (11.11) gives:

$$I_C = I_{CB0} - b\gamma I_E \qquad (11.12)$$

The leakage current I_{CB0} is small compared to the other two, so we may approximate:

$$I_C \simeq -b\gamma I_E \qquad (11.13)$$

Both b and γ are positive numbers between zero and one. In a good transistor they should both be close to unity, so a first order understanding of the transistor is:

$$I_C \simeq -I_E \qquad (11.14)$$

and hence from equation (11.8):

$$I_B \simeq 0 \qquad (11.15)$$

This means that the main current flow is between emitter and collector and the base current is small in comparison to either.

11.3.5 BJT circuit parameters

The *current gain* (or more strictly the common base current gain) α is defined:

$$\alpha = \frac{-\Delta I_C}{\Delta I_E} = b\gamma \qquad (11.16)$$

and can be thought of as an overall efficiency. (α is defined as a slope quantity – change in I_C/change in I_E – *not* as the ratio $-I_C/I_E$. It arises by differentiating equation (11.12) with respect to I_E, taking $I_{CB0} \simeq$ constant. The derivative and ratio *quantities* are very nearly the same because I_{CB0} is small, but the meanings are quite different.) It is typically 0.99 in a good device. This means that if the emitter current is increased by 100 mA the collector current changes by 99 mA, and hence the base current changes by 1 mA. Alternatively, we can say that if the base current changes by 1 mA, the collector current changes by 99 mA – and this begins to look like a high gain device.

To explore this idea more fully, we must return to equations (11.8)–(11.11) and derive an expression relating collector current to base current. Eliminate F_{eEB} between equations (11.9) and (11.10):

$$I_B + I_{CB0} = I_E(b\gamma - 1) \qquad (11.17)$$

Now substitute for I_E between equations (11.12) and (11.17):

$$I_B + I_{CB0} = \frac{(b\gamma - 1)}{b\gamma} (I_{CB0} - I_C) \qquad (11.18)$$

or:

$$I_C = \frac{\alpha}{1 - \alpha} I_B + \frac{I_{CB0}}{1 - \alpha} \tag{11.19}$$

The term $\alpha/(1 - \alpha)$ has a typical value of 100, and is given the symbol β or h_{FE} in many texts. This is called the common emitter current gain. The current, $I_{CB0}/(1 - \alpha)$, is the collector current which flows when the base is open circuited ($I_B = 0$), and is often given the symbol I_{CE0}. Equation (11.19) can then be written:

$$I_C = \beta I_B + I_{CE0} \tag{11.20}$$

A change ΔI_B in the base current will therefore cause a much larger change $\beta \Delta I_B$ in the collector current. This is the basis of amplification in the bipolar junction transistor.

11.4 Charge control model

The analysis of currents in the last section shows the link between the base and collector currents but it does not necessarily make clear which is cause and which is effect. It could even be argued that the BJT is a useless amplifier because I_C is always lower than I_E, for example. To suggest the train of events which give amplication we consider the circuit in Figure 11.8. The input is a small oscillating voltage applied through the capacitor, C, which blocks dc current flow. The output point is the junction between resistor $R3$ and the collector.

First we will show that this circuit can give the normal bias condition: base–emitter forward biased and collector–base reverse biased. This requires a

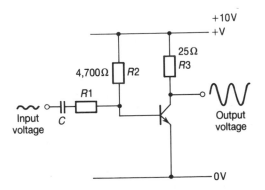

Figure 11.8 Common emitter amplifier

dc analysis of the currents in the circuit, *assuming* the normal bias condition exists. If a self-consistent picture is obtained, then the assumption is correct. A forward biased B–E junction will fix the base potential to one diode drop above the emitter potential, so $V_B = +0.7$ V. The current through $R2$ which flows into the base is then:

$$I_B = (10 - 0.7)/4{,}700 = 2 \text{ mA} \tag{11.21}$$

If $\beta = 100$ and $I_{CE0} \simeq 0$ this makes the collector current:

$$I_C \simeq \beta I_B = 200 \text{ mA} \tag{11.22}$$

and the collector potential V_C is:

$$V_C = 10 - (0.2 \times 25) = 5 \text{ V} \tag{11.23}$$

The collector is thus 4.3 V positive with respect to the base, which in an *npn* transistor makes the collector–base junction reverse biased. The dc analysis agrees with the initial assumption.

Now suppose the oscillating input voltage is applied. It is no good connecting it straight to the base because the base potential can vary only very slightly from 0.7 V. The resistor $R1$ converts the input voltage change ΔV to an input current change $\Delta I_B = \Delta V/R1$. This should not exceed the steady value of 2 mA calculated above.

The purpose of introducing this example circuit is to suggest how a variable base current could be applied to a normally biased BJT. Now we delve inside the device to see how a change in base current can cause a much larger change in the collector current. The key is to realize that the bulk of the base region, outside of junction depletion regions, is electrically *neutral*. The *p*-type material is normally neutral but there are two additional charge populations present when the transistor is biased. These are:

1. $\hat{Q}_{eB}$: the excess electron charge in the base injected from the emitter.
2. $\hat{Q}_{hB}$: the excess hole charge in the base due to the base current.

Taking both these quantities as positive, neutrality means:

$$\hat{Q}_{eB} = \hat{Q}_{hB} \tag{11.24}$$

The lifetimes of an electron and a hole in the base are not the same because they have different loss mechanisms. Holes can either be injected into the emitter or else recombine with an electron, and in a good transistor the second route dominates. The hole lifetime in the base is therefore the minority carrier lifetime, τ_e in an *npn* (Figure 11.9). Electrons injected from the emitter are nearly all gathered by the collector, so their lifetime in the base is typically their transit time τ_t. Thus the excess hole charge in the base is replaced every τ_e seconds by the base current:

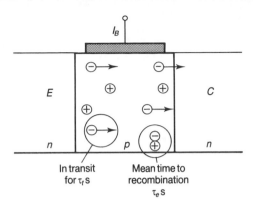

Note:
The neutral base region has equal numbers of injected electrons from the emitter and <u>extra</u> holes from the base.

Figure 11.9 Charge control model

$$I_B = \hat{Q}_{hB}/\tau_e \tag{11.25}$$

and the excess electron charge in the base is collected every τ_t seconds:

$$I_C = \hat{Q}_{eB}/\tau_t \tag{11.26}$$

Neglecting I_{CE0} the ratio is therefore:

$$\beta \simeq \frac{I_C}{I_B} = \frac{\hat{Q}_{eB}}{\hat{Q}_{hB}} \frac{\tau_e}{\tau_t} = \frac{\tau_e}{\tau_t} \tag{11.27}$$

This incidently gives a physical meaning to β and suggests how a good BJT should be made. τ_e should be long, so the base needs to be free of recombination centres. τ_t should be short, so the base region should be narrow.

Now suppose I_B is increased (Figure 11.10):

1. The hole current into the base rises.
2. $\hat{Q}_{hB}$ increases and the base becomes slightly positively charged.
3. This reduces the emitter–base barrier height.
4. More electrons diffuse from the emitter to the base.
5. The collector current increases.

The rise in I_B causes a temporary imbalance in charge flows, which alters the amount of charge in the depletion regions. This adjusts the barrier potentials to restore a new steady state condition. The only proviso is that the transistor should be able to respond quickly to changes in base current: this will be examined in more detail in the next chapter.

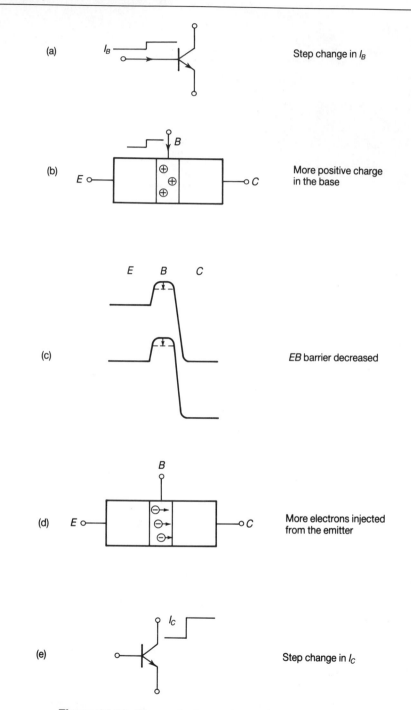

(a) I_B Step change in I_B

(b) E More positive charge in the base

(c)   *EB* barrier decreased

(d) E More electrons injected from the emitter

(e)   Step change in I_C

Figure 11.10 Change in I_B causing a change in I_C

11.5 Design of the base–emitter diode – BJT efficiency

The size and doping constraints on the base and emitter will be considered in this section. The quantities to be defined are (Figure 11.11).

1. W_E: emitter width normal to the junction.
2. W_B: base width normal to the junction.
3. N_{DE}: emitter doping number density (*npn* device).
4. N_{AB}: base doping number density (*npn* device).

The values of these strongly influence the gain parameters α and β. It will be necessary to repeat the analysis of the forward biased diode briefly. We recall that an applied forward bias V_A volts increases the minority carrier densities on each side of the depletion region by a factor $\exp(qV_A/kT)$. These excess minority carriers diffuse away from the junction until they recombine or reach the far side of the diode. The diffusion equation for $\hat{n}$ excess electrons in the *p*-type base is:

$$\frac{d^2\hat{n}}{dx^2} - \frac{\hat{n}}{D_e\tau_e} = 0 \tag{11.28}$$

In a very short base region recombination could be completely neglected, which

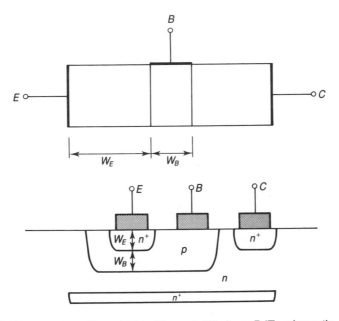

Figure 11.11 Base and emitter widths W_B and W_E in a BJT schematic and a planar device

is equivalent to letting $\tau_e \to \infty$ in equation (11.28), so that the second term vanishes. The equation can be integrated twice with the boundary conditions:

$$x = 0, \hat{n} = \frac{n_i^2}{N_{AB}} \left(\exp \frac{qV_A}{kT} - 1 \right) \tag{11.29}$$

$$x = W_B, \hat{n} = 0 \tag{11.30}$$

The minus one in the first expression is because $\hat{n}$ is the excess number density, not the total number density. The solution is:

$$\hat{n} = \frac{n_i^2}{N_{AB}} \left(1 - \frac{x}{W_B} \right) \left(\exp \frac{qV_A}{kT} - 1 \right) \tag{11.31}$$

The diffusion current of electrons injected into the base at $x = 0$ is therefore:

$$qF_{eEB} = -qAD_e \frac{d\hat{n}}{dx} \bigg|_{x=0} = \frac{qAD_e n_i^2}{N_{AB} W_B} \left(\exp \frac{qV_A}{kT} - 1 \right) \tag{11.32}$$

where A is the diode cross-sectional area. Similarly the hole current injected from the base into the emitter is:

$$qF_{hBE} = \frac{qAD_h n_i^2}{N_{DE} W_E} \left(\exp \frac{qV_A}{kT} - 1 \right) \tag{11.33}$$

This assumes that the emitter width W_E is much smaller than the hole diffusion length L_h. Equation (11.6) can now be used to give an expression for the emitter injection efficiency:

$$\gamma = \frac{D_e N_{DE} W_E}{D_e N_{DE} W_E + D_h N_{AB} W_B} \tag{11.34}$$

Thus $\gamma \to 1$ if $D_e N_{DE} W_E \gg D_h N_{AB} W_B$. The diffusion coefficients and widths tend to be of the same order in the base and emitter, so a high emitter injection efficiency requires:

$$N_{DE} \gg N_{AB} \quad (npn) \tag{11.35}$$

This coincides with the intuitive idea that heavy emitter doping will produce a larger diffusion current from the emitter when the junction is forward biased. The emitter is closer to the semiconductor surface in most integrated BJTs because it is easiest to make the doping density highest at the surface, and lower in the bulk.

To gain quantitative information on the base transport factor b, it is necessary to tighten up the analysis and allow for a finite value of τ_e: i.e. some electrons recombine in transit through the base. A general solution to the diffusion equation (11.28) is:

$$\hat{n} = C_1 \exp x/L_e + C_2 \exp -x/L_e \tag{11.36}$$

where C_1 and C_2 are constants determined by the boundary conditions and L_e is the minority carrier diffusion length in the base. When the boundary conditions equations (11.29) and (11.30) are inserted the expressions obtained contain hyperbolic trigonometric functions. That for $\hat{n}$ is:

$$n = \frac{n_i^2 \sinh\left[(W_B - x)/L_e\right]}{N_{AB} \sinh\left(W_B/L_e\right)} \left(\exp\frac{qV_A}{kT} - 1\right) \tag{11.37}$$

If this is differentiated with respect to x, the gradient can be evaluated at $x = 0$ and $x = W_B$. The ratio of these gradients is also the ratio of the diffusion currents at the base–emitter junctions and the base–collector junction, which is none other than the base transport factor. (Allow about an hour and two cups of coffee to check the next result yourself.) b turns out to be given by:

$$b = \operatorname{sech}(W_B/L_e) \tag{11.38}$$

A series expansion for $\operatorname{sech}(z)$ is $1 - z^2/2 + 5z^4/24 \dots$ so for $W_B \ll L_e$, the short base approximation is:

$$b \simeq 1 - \frac{W_B^2}{2L_e^2} \tag{11.39}$$

Thus for a base transport factor of 0.99 with a diffusion length of 20 microns, the base width should be 2.8 microns wide. This kind of definition is far easier to achieve vertically, by the difference in penetration distances of doping profiles, than horizontally, by photolithography. Thus lateral bipolar transistors will rarely have the gain of integrated versions with the active base and collector vertically below the emitter.

This last result contains the fact that *npn* transistors have higher gains than similar *pnp* transistors, because D_e is larger than D_h and hence L_e is larger than L_h. This is accentuated in graded base transistors, in which the base doping is non-uniform. This creates a built in electric field which reinforces the diffusion of injected minority carriers towards the collector and decreases the transit time across the base.

EXAMPLE

■ The base of an *npn* bipolar transistor is uniformly doped 10^{22} m^{-3}. Choose a base width and an emitter doping such that $\alpha = 0.98$, given that $\mu_e = 0.14$ m^2 V^{-1}s^{-1}, $\mu_h = 0.05$ m^2 V^{-1}s^{-1}, $\tau_e = \tau_h = 20$ μs, $T = 300$ K, $W_E = 2$ μm.

☐ There is no unique solution to this problem because α is a function of both N_{DE} and W_B, neither of which is fixed. One possible route is to divide the inefficiencies equally between γ and b, such that $b = \gamma = \sqrt{\alpha} = 0.99$.

The electron and hole diffusion lengths are given by:

$$L_e = (D_e \tau_e)^{1/2} = (\mu_e k T \tau_e / q)^{1/2} = 2.69 \times 10^{-4} \text{ m} \tag{11.40}$$

$$L_h = (D_h \tau_n)^{1/2} = (\mu_h k T \tau_h / q)^{1/2} = 1.61 \times 10^{-4} \text{ m} \tag{11.41}$$

If $b = 0.99$, then:

$$W_B = (2L_e^2 (1 - b))^{1/2} = 3.79 \times 10^{-4} \text{ m} \tag{11.42}$$

Rearranging equation (11.34):

$$N_{DE} = \gamma D_h N_{AB} W_B / D_e W_E (1 - \gamma) \tag{11.43}$$

The Einstein relation gives:

$$D_e = \mu_e k T / q = 3.61 \times 10^{-3} \text{ m}^2 \text{s}^{-1} \tag{11.44}$$

$$D_h = \mu_h k T / q = 1.29 \times 10^{-3} \text{ m}^2 \text{s}^{-1} \tag{11.45}$$

Hence if $W_B = 379 \ \mu\text{m}$, $W_E = 2 \ \mu\text{m}$, and $\gamma = 0.99$:

$$N_{DE} = 6.7 \times 10^{25} \text{ m}^{-3} \tag{11.46}$$

This is close to the practical upper limit for doping, while the value calculated for the base width is quite large. Any further increase in α could best be provided by narrowing the base.

11.6 Design of the base–collector diode – BJT breakdown

The emitter–base diode is the dominant influence on the gain of a BJT, but the base–collector diode is still important. In normal operation it determines the breakdown point of the transistor. The collector width does not matter so much for two reasons:

1. The construction of a planar BJT means that the metal contact is distant and the diode is wide.
2. The collector receives majority carriers, so the recombination and diffusion of minority carriers are not relevant in normal operation, e.g. a *pnp* collector receives holes which diffuse through the base.

The collector doping (N_{DC} in an *npn* device) does matter. Breakdown in a BJT is like that in a reverse biased diode and will occur in the base–collector diode first with normal biasing. The three main paths to breakdown are:

1. *Punchthrough* (Figure 11.12): the depletion region of the base–collector diode extends a distance d_p into the base in an *npn* device, given by

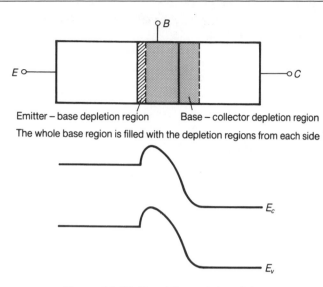

Figure 11.12 Punchthrough breakdown

equation (6.46):

$$d_p = \left(\frac{2\epsilon V_{BC}}{q} \frac{N_{DC}}{N_{AB}(N_{AB} + N_{DC})} \right)^{1/2} \qquad (11.47)$$

For example, if $N_{AB} = N_{DC} = 10^{22}$ m^{-3} and $V_{BC} = 100$ V in a silicon BJT, $d_p = 2.6\ \mu$m. This could easily extend all the way across the base and join the n^+ emitter directly to the n-type collector, causing a large current to flow. This is punchthrough breakdown. To avoid it, consider the ratio d_p/d_n of depletion region widths in the base and collector of an *npn* device (from equation (6.40), the charge neutrality condition):

$$\frac{d_p}{d_n} = \frac{N_{DC}}{N_{AB}} \qquad (11.48)$$

If $N_{AB} \gg N_{DC}$, $d_p \ll d_n$ and most of the depletion region is within the collector. In this case d_p becomes:

$$d_p \simeq \left(\frac{2\epsilon V_{BC} N_{DC}}{q N_{AB}{}^2} \right)^{1/2} \qquad (11.49)$$

Then if $N_{AB} = 10^{23}$, $N_{DC} = 10^{22}$ m^{-3} and $V_{BC} = 100$ V, d_p is $0.36\ \mu$m. Making most of the depletion region fall within the collector has another beneficial effect: *changes* in the depletion region width also occur mostly inside the collector. Strictly, the active base width is between the edges of the two depletion regions. A change in biasing, due even to a signal being amplified, will alter the base width and hence alter the gain. This is known as the Early effect. Making the base doping heavier than the collector doping minimizes this effect.

2.   *Avalanche*: this is the creation of a rush of extra electron–hole pairs by charges gaining energy in a potential gradient. The maximum electric field in a BJT will occur at the metallurgical junction of base and collector in normal bias, because this is the reverse biased junction. The field in an *npn* transistor will be (equation (6.49)):

$$\mathscr{E}_{max} = \left(\frac{2qV_{BC}N_{AB}N_{DC}}{\epsilon(N_{AB} + N_{DC})}\right)^{1/2} \tag{11.50}$$

$$\approx \left(\frac{2qV_{BC}N_{DC}}{\epsilon}\right)^{1/2}$$

if $N_{AB} \gg N_{DC}$. Breakdown occurs if $\mathscr{E}_{max}$ exceeds the avalanche value for the material. Thus N_{DC} should be kept small to keep the $\mathscr{E}_{max}$ small. However, too low a value will lead to an excessive collector resistance.

3.   *Excess power*: the band diagram for a BJT in normal operation (Figure 11.13) shows that base minority carriers which cross into the collector lose several electron-volts of potential energy. This energy is rapidly removed from the mobile charges and appears as heat. The current in a single reverse biased *pn* diode is so small that the heating is negligible; but this is not the case in a transistor. Injection of minority carriers into the base by the emitter means that there is now a substantial current flowing in the reverse biased base–collector diode. The power dissipation P must not exceed P_{max} for the device:

$$P = I_C V_{CE} \leq P_{max} \tag{11.51}$$

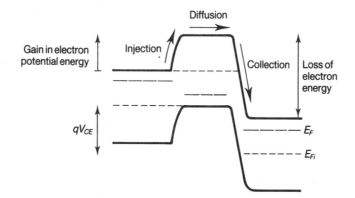

Figure 11.13 Electron injection cools the emitter–base junction region and heats the base–collector junction region

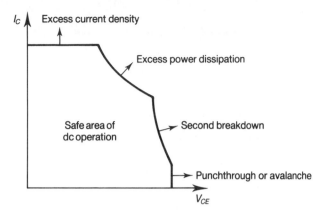

Figure 11.14 Safe area of BJT operation

This defines one boundary of the safe area of operation of a BJT, as illustrated in Figure 11.14. A further restriction can appear at the high voltage end due to hot spots developing inside the device. Heating will cause the intrinsic electron number density n_i to rise and the resistivity to fall, which exacerbates the current crowding effect. Melting and catastrophic destruction will follow well before the unfortunate engineer can reach for the off switch. This phenomenon is called second breakdown, because it follows a non-destructive avalanche breakdown.

11.7 The Ebers–Moll model

The discussion has so far been limited to normal transistor action biasing. Switching and high frequency behaviour will be treated in the next chapter and a model will be given in this section which can treat both normal and inverse transistor action. The model has two main uses:

1. It gives a mathematical representation of a transistor which works even if the emitter and collector are interchanged.
2. It forms the basis for more complex models suitable for computer simulation of transistor circuits, when normal biasing cannot be assumed.

Consider first a normal bias *pnp* transistor (Figure 11.15a). A current I_{E1} flows through the emitter–base diode, given by the diode equation:

$$I_{E1} = I_{E0} \left(\exp \frac{(qV_{EB})}{kT} - 1 \right) \tag{11.52}$$

where I_{E0} is the reverse saturation current for this diode. A current $\alpha_F I_{E1}$ flows

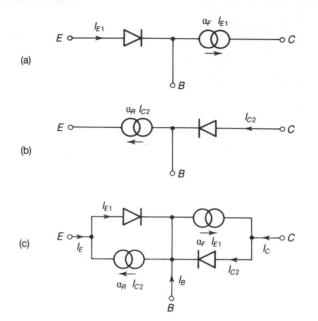

Figure 11.15 The Ebers–Moll BJT model

into the collector (subscript F = forward operation), represented by the current generator. The condition in reverse operation is shown in Figure 11.15b, again for a *pnp* device, as indicated by the sense of the diode. The collector–base diode current is:

$$I_{C2} = I_{C0}\left(\exp\frac{qV_{CB}}{kT} - 1\right)$$ (11.53)

where I_{C0} is the collector–base diode reverse saturation current. The current $\alpha_R I_{C2}$ flows in the emitter, operating as a collector in the reverse mode. The two can be superimposed as in Figure 11.15c, to give a model, first proposed by Ebers and Moll, which works in either bias condition. The current equations are:

$$I_E = I_{E0}\left(\exp\frac{qV_{EB}}{kT} - 1\right) - \alpha_R I_{C0}\left(\exp\frac{qV_{CB}}{kT} - 1\right)$$ (11.54)

$$I_C = -\alpha_F I_{E0}\left(\exp\frac{qV_{EB}}{kT} - 1\right) + I_{C0}\left(\exp\frac{qV_{CB}}{kT} - 1\right)$$ (11.55)

$$I_B = -I_E - I_C$$ (11.56)

These apply to the dc or low frequency operation of a *pnp* bipolar junction transistor. At first sight four parameters are needed to characterize the transistor: α_F, α_R, I_{E0} and I_{C0}, with the voltages being defined by the total circuit. In fact a further relationship can be deduced, either by reciprocity arguments or by deriving expressions for the four parameters from the charge diffusion equation. It can be shown that:

$$\alpha_R I_{C0} = \alpha_F I_{E0} \tag{11.57}$$

which means that a simple transistor model requires only three parameters.

EXAMPLE 1

■ The base transport factor of a *pnp* transistor is 0.97: the base width is 20 μm, $\mu_h = 0.19$ m^2 V^{-1} s^{-1} and T = 300 K. Find the hole lifetime in the base.

☐ The base transport factor for a *pnp* transistor, from equation (11.39):

$$b = 1 - \frac{W_B{}^2}{2L_h{}^2}$$

so:

$$L_h = \left[\frac{W_B{}^2}{2(1-b)}\right]^{1/2} = 8.16 \times 10^{-5} \text{ m}$$

Now the hole diffusion length is given by:

$$L_h = (D_h \tau_h)^{1/2}$$

Using the Einstein relation:

$$\tau_h = L_h{}^2/D_h = \frac{qL_h{}^2}{\mu_h kT}$$

$$= 1.36 \times 10^{-5} \text{ s}$$

EXAMPLE 2

■ An *npn* transistor has emitter, base and collector doping densities of 10^{25}, 10^{23} and 10^{21} m^{-3}. The emitter and base widths are 3 μm and 5 μm respectively; $L_e = 10$ μm and $L_h = 6$ μm. Find the forward and reverse common base current gains if $\tau_e = \tau_h$.

☐ The base transport factor is the same in either case. The more accurate expression equation (11.30) can be used:

$$b = \text{sech}(W_B/L_e)$$
$$= 0.887$$

(If the approximation equation (11.39) is used, $b = 0.875$.)

Equation (11.34) gives the emitter injection efficiency γ. The diffusion coefficients are not known, so γ must be expressed in terms of the diffusion lengths: $D_{e,h} = L_{e,h}{}^2/\tau_{e,h}$:

$$\gamma = \frac{L_e{}^2 N_{DC} W_E}{D_e{}^2 N_{DE} W_E + L_h{}^2 N_{AB} W_B}$$

τ_e and τ_h are equal and cancel out. This can be evaluated for the forward case:

$$\gamma_F = 0.943$$

Hence:

$$\alpha_F = b\gamma_F = 0.837$$

In reverse operation N_{DC} replaces N_{DE} in the expression for γ. The collector is not a short diode, so W_E is replaced by L_e:

$$\gamma_R = \frac{L_e{}^3 N_{DC}}{L_e{}^3 N_{DC} + L_h{}^2 N_{AB} W_B}$$
$$= 0.053$$

Then:

$$\alpha_R = b\gamma_R = 0.047$$

When the base doping is much larger than the collector doping, α_R is very low.

EXAMPLE 3

■ Find the punchthrough breakdown voltages for the transistor in (2) in forward and reverse operation. Take $\epsilon_r = 11.9$.

□ We use equation (11.47) with $d_p = W_B$, the punchthrough condition. Rearranging in terms of the base–collector voltage:

$$V_{BC} = \frac{q N_{AB}(N_{AB} + N_{DC}) W_B{}^2}{2\epsilon N_{DC}}$$

In forward operation, with $N_{AB} = 10^{23}$ m^{-3}, $N_{DC} = 10^{21}$ m^{-3}, and $W_B = 5$ μm:

$$V_{BC} = 1.92 \times 10^5 \text{ V}$$

This is extremely high – the BJT will almost certainly fail by avalanche at a lower voltage in forward operation. In reverse operation the emitter doping replaces the collector doping in the expression and V_{BC} becomes V_{BE}. At punchthrough:

$$V_{BE} = \frac{qN_{AB}(N_{AB} + N_{DE})W_B{}^2}{2\epsilon N_{DE}}$$

with $N_{DE} = 10^{25} \text{ m}^{-3}$:

$$V_{BE} = 19.2 \text{ V}$$

When the base doping is lighter than the collector doping punchthrough occurs very easily.

Summary

Main symbols introduced

Symbol	Unit	Description
b		base transport factor
F	charges s^{-1}	charge flow
α		common base current gain
β		common emitter current gain
γ		emitter injection efficiency
τ_t	s	base transit time

Subscripts

B	base
C	collector
E	emitter
F	forward operation
$L, 0$	leakage value or reverse saturation value
R	reverse operation

Main equations

$$I_E + I_B + I_C = 0 \quad \text{(currents flowing into BJT positive)} \tag{11.1}$$

$$\gamma = F_{eEB}/(F_{eEB} + F_{hBE}) \tag{11.6}$$

$$= D_e N_{DE} W_E/(D_e N_{DE} W_E + D_h N_{AB} W_B) \quad (npn) \tag{11.34}$$

$$b = F_{eBC}/F_{eEB} \tag{11.7}$$

$$\simeq 1 - (W_B{}^2/2L_e{}^2) \quad (npn) \tag{11.39}$$

$$\alpha = -\Delta I_C/\Delta I_E = b\gamma \tag{11.16}$$

$$\beta = \alpha/(1 - \alpha) \tag{11.19}$$

$$= \tau_e/\tau_t \tag{11.27}$$

$$I_C V_{CE} \le P_{max} \tag{11.51}$$

Exercises

1.   Outline the fabrication sequence used to make a lateral *pnp* transistor like that in Figure 11.4.

2.   Suggest how the transistor in Figure 11.16 could be made. This structure is called collector diffusion isolation because it does away with the separate isolation ring around each transistor. Can *pnp* transistors be made at the same time?

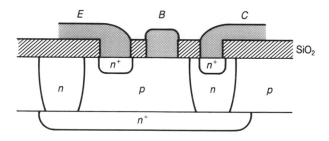

Figure 11.16 A collector diffusion isolation structure

3.   Draw the diagram for charge flows in a *pnp* transistor corresponding to Figure 11.6 and write down the Kirchoff current equations (11.1)–(11.4) for the *pnp* case.

4.   Show that $W_B{}^2/D_e$ has the units of time. If this is an approximation to the base transit time in an *npn* transistor, find β for a transistor where $W_B = 25\ \mu m$, $\tau_e = 20\ \mu s$, $\mu_e = 0.135\ m^2\,V^{-1}s^{-1}$ and $T = 300$ K. How could β be increased in this device?

5. Draw the band diagrams for a p^+np transistor in forward and reverse operation, paying attention to the depletion regions. Find the current through the p^+ region in both cases if $I_B = -1$ mA, $\alpha_F = 0.9$ and $\alpha_R = 0.3$.

6. An *npn* BJT has an emitter injection efficiency of 0.99 and a base transport factor of 0.96. The base–collector reverse saturation current is 2 μA. Find the common emitter current gain and the collector current for a base current of 5 mA.

7. Choose a base and emitter doping for a silicon *npn* bipolar transistor such that $\beta = 70$, given that $\mu_e = 0.135$, $\mu_h = 0.048$ m^2 V^{-1} s^{-1}, $\tau_e = \tau_h = 30$ μs, $W_E = 5$ μm, $W_B = 3$ μm and $T = 300$ K. (There is no unique solution.)

8. A *pnp* transistor has emitter, base and collector widths of 2.5 μm, 5 μm and 20 μm respectively. The emitter, base and collector doping levels are 10^{24}, 10^{23} and 10^{22} m^{-3}, and the mobilities and lifetimes are as in (7). Find the forward and reverse values of α.

9. Find the punchthrough and avalanche breakdown voltages for the device in (8) in both forward and reverse operations. Take $\epsilon_r = 11.9$ and the avalanche field as 3×10^7 V m^{-1}.

10. Use the Ebers–Moll model to find the base current which flows with the base-emitter junction forward biased 0.4 V and the other diode reverse biased 5 V, if $\alpha_F = 0.98$, $\alpha_R = 0.1$ and $I_{C0} = 1$ μA. Find the common emitter current gain and the power dissipation in the transistor.

12 Bipolar junction transistors – ac

12.1 The I/V characteristic

In the last chapter the basic principles of transistor action were outlined: the bipolar junction transistor is a current-controlled device, which is normally operated with the base–emitter junction forward biased and the base–collector junction reverse biased. Carriers injected into the base from the emitter mostly diffuse across the base into the collector, while those lost by recombination constitute the largest part of the base current.

There are two main ways of looking at which variables are the inputs and which are the outputs in normal operation. The voltages V_{BE} and V_{CB} are rarely considered as input variables because their main role is to set up the bias condition. Once the base–emitter diode is turned on, V_{BE} is pinned to about 0.7 V in silicon, regardless of the diode current. It would be very difficult to control the BJT by making tiny adjustments to V_{BE} (Figure 12.1a). Likewise, once the base–collector junction is reverse biased it will collect all minority carriers reaching the depletion region edge in the base (Figure 12.1b). Changes in V_{BC} will only alter the base width and the reverse saturation current I_{CB0} a little, so the collector current is independent of V_{BC} over quite a wide range. Neither V_{BE} nor V_{BC} are good input variables in normal operation.

The currents I_E and I_B are the parameters usually taken as input variables. The collector current could also be taken as an input, but is more often the

242

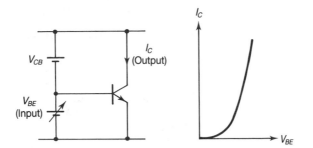

(a) Operated with variable V_{BE} as input signal

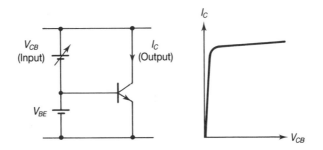

(b) Operated with variable V_{CB} as input signal

Figure 12.1 Bipolar transistor circuit

output variable. We have already used the terms α, the common base current gain and β, the common emitter current gain. Now the meaning will be made explicit.

Look first at Figure 12.2a. The input is now a current generator in the emitter lead, depicted as a double circle device. This is presumed to be capable of fixing I_E to any desired value. (It could supply all the voltage bias necessary, but the bias voltage battery has been left as a reminder that the junction is forward biased.) It is clear that there are two main current loops, one involving each junction, and the base is common to each. The input/output characteristic (Figure 12.2b) is almost linear with a slope of $-\alpha$. The minus sign arises because we have taken inward flowing currents as positive.

The common base configuration does not supply current gain because α is always less than unity. However, it can provide power gain. The power dissipation in the input loop is $|I_E V_{BE}|$, while that in the output loop is $|I_C V_{BC}|$. V_{BE} is constrained by the forward bias turn on voltage but V_{BC} is limited by the reverse breakdown voltage. Since $|I_C| \simeq |I_E|$, this means that small changes in input power can control much larger changes in output power.

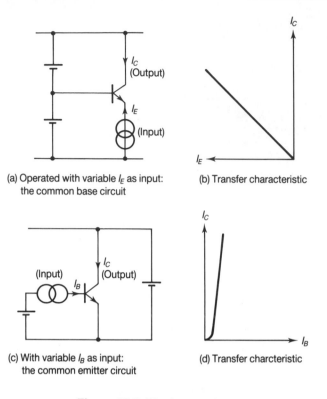

(a) Operated with variable I_E as input: the common base circuit

(b) Transfer characteristic

(c) With variable I_B as input: the common emitter circuit

(d) Transfer charcteristic

Figure 12.2 Bipolar transistor

The common emitter configuration is shown in Figure 12.2c. Again there are two main current loops, each containing the emitter this time. The base current is controlled by a current generator, which in turn determines the collector current (Figure 12.2d). Small changes ΔI_B cause much larger changes $\beta \Delta I_B$ in I_C, so there is current amplification in this case.

The overall I/V characteristics are usually drawn as in Figure 12.3a and b. The collector current is the output in both cases. The voltage across the transistor in the output loop is used on the voltage axis: V_{CB} in the common base circuit and V_{CE} in the common emitter case. The input currents are treated as a third parameter, generating a family of curves. It should be understood that the curves drawn are a selection from an infinite number of curves for all values of input current.

In most circuits there is a load resistance R in series with the collector, so that a change in I_C also causes a change in the collector voltage, because of the altered potential across R. Suppose the common emitter circuit in Fig 12.4 is driven by a constant voltage V and a resistance R is included in series with the

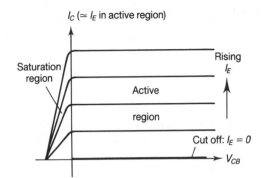

I_C ($\simeq I_E$ in active region)

Saturation
region

Active

region

Rising
I_E

Cut off: $I_E = 0$

V_{CB}

(a) *npn* common base output characteristic

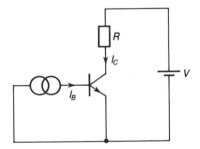

I_C ($\simeq \beta I_B$ in active region)

V/R

Saturation

Rising
I_B

Active region

V_{CE}

Cut off: $I_B = 0$ V

(b) *npn* common emitter output characteristic, including load-line
   $V_{CE} + I_C R = V$

Figure 12.3 Overall *I/V* characteristics

R

I_C

V

I_B

Note:
The transistor voltage V_{CE} now depends on I_C.

Figure 12.4 Common emitter circuit with a load resistance *R*

collector. The relationship between V_{CE} and I_C for the circuit can then be written:

$$V = V_{CE} + I_C R \text{ V} \tag{12.1}$$

This is another example of a load-line, and can be plotted on the same axes as the common emitter characteristic, as in Fig. 12.3b. When run as a small signal amplifier the circuit supplies a steady value for I_B to make the operating point sit in the middle of the region in which transistor action occurs, known also as the active region. Small changes ΔI_B in I_B are equivalent to moving from one curve to another in the $I_C V_{CE}$ graph. The circuit responds by shifting up and down the load line as both I_C and V_{CE} alter to fit in with the new values of base current.

Next consider the limits of the active region. As I_B is reduced to zero the collector current also falls to a low value and most of the applied potential appears across the transistor. This is called *cut off*, and is equivalent to losing forward bias on the emitter–base diode. At the other extreme of the load-line a high base current causes a large collector current to flow while V_{CE} drops to a low value, typically 0.2 V. Here the base–collector diode is no longer reverse biased, a condition called *saturation*. (Saturation must be one of the most heavily over-used words in engineering.) Notice that V_{CE} can fall *below* the switch on value of V_{BE} because the voltages sum in opposition (Figure 12.5a):

$$V_{CE} = V_{CB} + V_{BE} \text{ V} \tag{12.2}$$

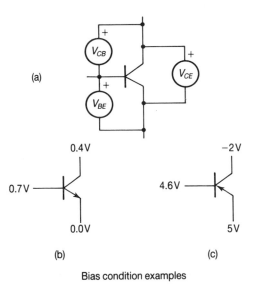

Bias condition examples

Figure 12.5 Definition of V_{CB}, V_{BE}, V_{CE} and bias condition examples

In an *npn* device in normal operation both V_{CB} and V_{BE} are positive. In saturation V_{CB} goes negative while V_{BE} remains positive, so $|V_{CE}|$ will be less than either $|V_{CB}|$ or $|V_{BE}|$.

EXAMPLE

■ The voltage measured with respect to ground on the emitter, base and collector of an *npn* and a *pnp* bipolar transistor are given in Figures 12.5b and c. Determine the state of each device.

□ 1. *npn*

$$V_C = 0.4 \text{ V} \quad V_B = 0.7 \text{ V} \quad V_E = 0 \text{ V}$$

$$V_{BE} = V_B - V_E = +0.7 \text{ V}$$

The *p*-type base is positive with respect to the *n*-type emitter, so this diode is forward biased:

$$V_{CB} = V_C - V_B = -0.3 \text{ V}$$

The *p*-type base is positive with respect to the *n*-type collector so this diode is forward biased. The device is in saturation.

2. *pnp*

$$V_C = -2 \text{ V} \quad V_B = 4.6 \text{ V} \quad V_E = 5 \text{ V}$$

$$V_{BE} = V_B - V_E = -0.4 \text{ V}$$

The *n*-type base is negative with respect to the *p*-type emitter, so the diode is forward biased:

$$V_{CB} = V_C - V_B = -6.6 \text{ V}$$

The *n*-type base is positive with respect to the *p*-type collector, so the diode is reverse biased. This transistor is in the active region.

12.2 Small-signal ac models

This section will concentrate on a bipolar junction transistor biased to be in the active region. The aim is to develop circuit models, in terms of resistance, capacitance and current or voltage sources, which display the same behaviour to *small ac signals* as does the transistor. As in the case of the single junction diode (section 7.1) it must be stressed that a small-signal equivalent circuit will not give the steady dc behaviour. However, it is often possible to divide the circuit behaviour into two distinct parts:

1. Dc: the currents and voltages which exist in the absence of an input signal.
2. Ac: the small changes from the steady dc values.

If the oscillations are not small, so that the transistor is driven into saturation or cut off, then another model which can deal with switching transients must be used.

The current gains α and β introduced in Chapter 11 are themselves small-signal quantities, defined by:

$$\alpha = -\left.\frac{\partial I_C}{\partial I_E}\right|_{V_{CB}=\text{constant}} \tag{12.3}$$

$$\beta = \left.\frac{\partial I_C}{\partial I_B}\right|_{V_{CE}=\text{constant}} \tag{12.4}$$

Suppose that the emitter and collector currents are composed of dc components I_E and I_C together with small ac signals i_e and i_c at a frequency ω (Figure 12.6):

$$I_E = I_E + i_e \sin(\omega t)\ \text{A} \tag{12.5}$$

$$I_C = I_C + i_c \sin(\omega t)\ \text{A} \tag{12.6}$$

The two currents are in phase provided ω is low enough to neglect capacitative currents. Differentiating with respect to time:

$$\frac{\partial I_E}{\partial t} = \omega i_e \cos(\omega t) \tag{12.7}$$

$$\frac{\partial I_C}{\partial t} = \omega i_c \cos(\omega t) \tag{12.8}$$

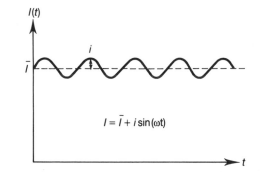

Figure 12.6 Total current $I(t)$, mean dc current $\bar{I}$ and ac current i

Hence:

$$\frac{\partial I_C}{\partial I_E} = \frac{\partial I_C}{\partial t} \bigg/ \frac{\partial I_E}{\partial t} = \frac{i_c}{i_e}\bigg|_{v_{cb}=0} \tag{12.9}$$

So α and β can be written in terms of the small-signal currents:

$$\alpha = -i_c/i_e|_{v_{cb}=0} \tag{12.10}$$

$$\beta = i_c/i_b|_{v_{ce}=0} \tag{12.11}$$

Be careful about these definitions, because there is no universal agreement on the use of symbols. In some literature α and β are the ratios of the *total* currents I_C/I_E and I_C/I_B. The numerical values are nearly the same, but the meanings are profoundly different.

12.2.1 Hybrid–π model

There are two ways of arriving at essentially the same equivalent circuit for a BJT connected in common emitter mode. The hybrid–π is based more closely on the physical construction of the transistor, while the h-parameter model presented subsequently is derived from circuit principles. The simplest version of the hybrid–π model is given in Figure 12.7. The small-signal base current i_b creates an ac voltage v_{be} across the input slope resistance r_{be}. The output current i_c is driven by an ac current source of strength $g_m v_{be}$, where g_m is the transconductance or mutual conductance defined by

$$g_m = \frac{\partial I_C}{\partial V_{BE}}\bigg|_{V_{CE}=\text{constant}} = \frac{i_c}{v_{be}} \tag{12.12}$$

This quantity has the units of 1/ohms or Siemens (S). It is called the *mutual* conductance because it expresses the change in an output current I_C due to a change in an input voltage V_{BE}. These equivalent circuit elements can be analysed as follows:

1. *Input resistance, r_{be}*: this is the slope resistance of the base–emitter diode, as seen by the ac *base* current:

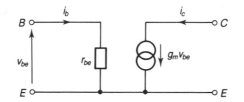

Figure 12.7 The simplest hybrid-π ac equivalent circuit for a BJT

$$r_{be} = \frac{\partial V_{BE}}{\partial I_B} = \frac{v_{be}}{i_b} \tag{12.13}$$

If $|i_c| \simeq |i_e|$ ($\alpha \simeq 1$) then $i_b \simeq i_e/\beta$ and

$$r_{be} \simeq \frac{\beta v_{be}}{i_e} \tag{12.14}$$

The ratio v_{be}/i_e is the slope resistance of the base–emitter diode as seen by the main *emitter* current, which according to section 7.1 is:

$$\frac{v_{be}}{i_e} = \frac{kT}{qI_E} \tag{12.15}$$

Hence:

$$r_{be} \simeq \frac{\beta kT}{qI_E} \tag{12.16}$$

For example, if $\beta = 100$, $kT/q = 26 \text{ mV}$ and $I_E = 50 \text{ mA}$ the input resistance is 52 Ω. Note that r_{be} is inversely proportional to the total emitter current.

2. *Mutual conductance* g_m: this was defined in equation (12.12). The diode equation for the base–emitter junction may be written:

$$|I_C| \simeq |I_E| \simeq I_{E0} \exp \frac{qV_{BE}}{kT} \tag{12.17}$$

Hence:

$$g_m = \frac{\partial I_C}{\partial V_{BE}} \simeq \frac{I_{E0}q}{kT} \exp \frac{qV_{BE}}{kT}$$

$$\simeq \frac{qI_E}{kT} \tag{12.18}$$

Thus g_m and r_{be} are related:

$$\beta = g_m r_{be} \tag{12.19}$$

A typical value for g_m is 2 S, at $I_E = 50 \text{ mA}$. Note that g_m is proportional to I_E, so β is independent of I_E to a first approximation.

Other small-signal elements can be added to this simple model, as in Figure 12.8. The three resistances are:

1. *Base spreading resistance* $r_{bb'}$: this is the only real resistance in the model, which accounts for the resistance of the semiconductor between the base contact and the active base volume. It is typically 10–50 Ω. The external base connection is designated b and the internal base b'. The input

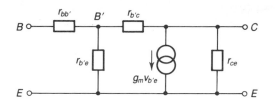

Figure 12.8 More accurate low-frequency hybrid-π circuit

resistance is renamed $r_{b'e}$ and the internal input voltage is $v_{b'e}$.

2. *Collector–base resistance $r_{b'c}$*: this element forms the horizontal bar of the π-shaped circuit and takes into account the variation in base width with collector voltage. Some portion of i_b flows to the common emitter point through $r_{b'c}$, lowering the current through $r_{b'e}$ and the potential $v_{b'e}$. A typical value of $r_{b'c}$ is 10 MΩ, indicating that the effect is small. The collector–base resistance is often replaced with an open circuit (i.e. omitted) in approximate analysis.

3. *Output resistance r_{ce}*: the collector current is not wholly independent of V_{CE} in the active region, but increases slowly as V_{CE} rises. This shows up on the common emitter I_C/V_{CE} characteristic curves as a small slope on the nominally flat parts of the lines. The physical origin of this is the Early effect, or variation in base width with bias potential. If the collector–base resistance can be omitted, then the collector current is:

$$i_c = g_m v_{b'e} + v_{ce}/r_{ce} \tag{12.20}$$

In a good transistor the effect will be small and r_{ce} will be large, typically 100 kΩ.

EXAMPLE

■ A BJT connected in common emitter mode has the following hybrid-π parameters: $r_{bb'} = 20\ \Omega$, $r_{b'e} = 100\ \Omega$, $g_m = 1.2\ S$, $r_{ce} = 120\ k\Omega$ and $r_{b'c}$ is high. Find the common emitter current gain, and the amplitudes of $v_{b'e}$ and i_c for an ac base current of 0.3 mA, if a 1 kΩ load is connected between collector and emitter, as in Figure 12.9.

☐ Using equation (12.19), modified to use the internal base:

$$\beta = g_m r_{b'e}$$
$$= 120$$

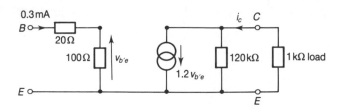

Figure 12.9 Example hybrid-π BJT equivalent circuit

The input base current flows through $r_{bb'}$ and $r_{b'e}$, so the potential across the internal input resistance is:

$$v_{b'e} = i_b r_{b'e} \tag{12.21}$$

$$= 0.03 \text{ V}$$

Note that this is small compared to the dc value of V_{BE}, satisfying the small-signal condition.

The small signal output voltage appears across the load resistance R_{Load}:

$$v_{ce} = i_c R_{Load} \tag{12.22}$$

This can be combined with equation (12.20) to give:

$$i_c = g_m v_{b'e} + i_c R_{Load}/r_{ce} \tag{12.23}$$

or

$$i_c = \frac{g_m v_{b'e}}{1 - (R_{Load}/r_{ce})} \tag{12.24}$$

$$= 36.3 \text{ mA}$$

Note that $i_c/i_b = 121$ which is slightly different from β. If $R_{Load} \ll r_{ce}$ then the common emitter current gain tends to β. The value of $r_{bb'}$ is irrelevant in this example because we have assumed that the BJT base is driven by a perfect current source. If the current i_b were derived from a voltage in series with a resistance, then $r_{bb'}$ would add to the external load resistance and make a difference to the behaviour obtained.

Finally, the hybrid–π model can be extended to include junction capacitances. Without this it cannot predict high frequency behaviour, because none of the small-signal circuit elements considered so far depend on the frequency of the oscillation. The full hybrid–π model is given in Figure 12.10, which contains two capacitors.

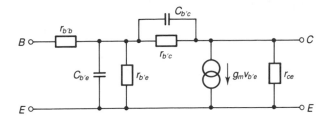

Figure 12.10 High-frequency hybrid-π BJT equivalent circuit

1. *Base–emitter capacitance $C_{b'e}$*: this is the sum of the diffusion capacitance, C_{diff} due to stored injected charge, and depletion capacitance C_d, caused by the charges in the depletion region. These were discussed section 7.1 and in forward bias the diffusion capacitance dominates. In this context it is defined in terms of the excess charge injected in the base $\hat{Q}_B$ ($\mathrm{C\,m^{-2}}$) and the emitter–base voltage:

$$C_{diff} = \frac{\partial \hat{Q}_B}{\partial V_{EB}} \; \mathrm{F\,m^{-2}} \tag{12.25}$$

If the depletion capacitance dominates $C_{b'e} \simeq A C_{diff}$ where A is the B–E diode cross-sectional area. Then:

$$C_{b'e} \simeq A \frac{\partial \hat{Q}_B}{\partial I_E} \frac{\partial I_E}{\partial V_{BE}} \; \mathrm{F} \tag{12.26}$$

According to equation (11.26), $I_E \simeq A\hat{Q}_B/\tau_t$, where τ_t is the transit time through the base. Using the diode equation for $\partial I_E/\partial V_{BE}$ we obtain:

$$C_{b'e} = \tau_t \; \frac{qI_E}{kT}$$

$$= \tau_t g_m \tag{12.27}$$

Thus for $\tau_t = 1$ ns, $I_E = 50$ mA, $C_{b'e} = 1.9$ nF.

2. *Collector–base capacitance, $C_{b'c}$*: The reverse biased base–collector junction has virtually no diffusion capacitance and only a small depletion capacitance. The increased width of the depletion region reduces the capacitance to a few picofarads.

The impedance of a capacitor is inversely proportional to frequency f for sinusoidal signals. Therefore at high frequency the input is almost short-circuited by $C_{b'e}$ and the base is shorted by $C_{b'c}$. The gain of the transistor will fall under these circumstances, so it is important to find the useful bandwidth of a BJT. The common emitter current gain is defined for $v_{ce} = 0$, which means no ac load

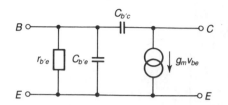

Figure 12.11 Approximate high-frequency hybrid-π circuit

resistance. In this case, neglecting $r_{b'c}$ and $r_{bb'}$ two capacitors appear in parallel (Figure 12.11) and the ac base current is:

$$i_b = v_{b'e} \left(\frac{1}{r_{b'e}} + j2\pi f C_{b'e} + j2\pi f C_{b'c} \right) \qquad (12.28)$$

while the ac collector current is:

$$i_c = g_m v_{b'e} \qquad (12.29)$$

Here j means the square root of minus one. Hence the common emitter current gain is:

$$\beta = \left. \frac{i_c}{i_b} \right|_{v_{ce}=0} = \frac{g_m r_{b'e}}{1 + j2\pi f (C_{b'e} + C_{b'c}) r_{b'e}} \qquad (12.30)$$

The transition from low frequency behaviour to high frequency behaviour occurs when the real and imaginary parts in the denominator are equal in magnitude:

$$f_\beta = \frac{1}{2\pi (C_{b'e} + C_{b'c}) r_{b'e}} \text{ Hz} \qquad (12.31)$$

Thus equation (12.30) can be rewritten:

$$\beta = \frac{g_m r_{b'e}}{1 + jf/f_\beta} \qquad (12.32)$$

At frequencies below f_β the gain is nearly constant, taking the value $g_m r_{b'e}$ derived before. When $f/f_\beta \gg 1$ we may approximate:

$$\beta \simeq -j g_m r_{b'e} f_\beta / f \quad (f \gg f_\beta) \qquad (12.33)$$

The two regimes are indicated in Figure 12.12, which is a graph of $\log \beta$ against $\log f$.

The $-j$ indicates that a 90° phase change occurs between i_b and i_c. Taking the magnitude only:

$$|\beta f| \simeq g_m r_{b'e} f_\beta = f_T \qquad (12.34)$$

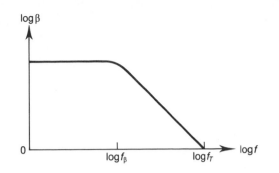

Figure 12.12 Log β against log frequency, based on the hybrid-π model

βf is the *gain–bandwidth product* for the transistor. This is also known as the transition frequency f_T. In the high frequency region above f_β the use of a BJT over a bigger frequency spread, or bandwidth, is accompanied by a loss in current gain. When choosing a particular transistor it is important to check that the gain will be sufficient at the highest frequency employed.

EXAMPLE

■   A transistor has a low frequency value of β of 80 with a dc collector current of 10 mA. The base transit time is 50 ns. Will it be suitable for an application requiring a common emitter current gain of 40 at 100 MHz?

☐   Using equation (12.18) for g_m, assuming $I_E \simeq I_C$:

$$g_m \simeq \frac{qI_E}{kT} \tag{12.35}$$

$$= 0.385 \text{ S}$$

at $T = 300$ K. Equation (12.19) gives $r_{b'e}$:

$$r_{b'e} = \beta/g_m \tag{12.36}$$

$$= 208 \ \Omega$$

The dominant base–emitter capacitance can also be calculated from equation (12.27):

$$C_{b'e} = \tau_t g_m \tag{12.37}$$

$$= 19.2 \text{ nF}$$

Hence the breakpoint frequency is, neglecting $C_{b'c}$ (equation (12.31)):

$$f_\beta \simeq 1/2\pi C_{b'e} r_{b'e} \tag{12.38}$$

$$= 125\ \text{kHz}$$

The specified 100 MHz is well above f_β, so the approximation equation (12.33) can be used:

$$|\beta| \simeq g_m r_{b'e} f_\beta / f$$

$$= 0.1$$

Current gain has disappeared completely at 100 MHz, so this transistor is unsuitable for the application specified.

12.2.2 h-parameter model

This will be developed for the low frequency case only, neglecting capacitances. The model treats the BJT as a sealed black box with four wires coming out of it. Two are labelled common, one is labelled input and one output. As for the hybrid-π model we will only consider the case where the emitter is the common terminal, and all the h-parameters will carry the subscript e. In the discussion below the ac input voltage and current are called v_1 and i_1, while the corresponding output quantities are v_2 and i_2. Four h-parameters are defined for the model circuit shown in Figure 12.13:

1. *Input resistance* h_{ie}: This is the ratio v_1/i_1 with $v_2 = 0$, i.e. with the output short-circuited:

    $$h_{ie} = \left.\frac{v_1}{i_1}\right|_{v_2=0} = \left.\frac{v_{be}}{i_b}\right|_{v_{ce}=0} \tag{12.39}$$

    in BJT terminology. The corresponding hybrid-π expression is:

    $$h_{ie} = r_{bb'} + \frac{r_{b'e} r_{b'c}}{r_{b'e} + r_{b'c}} \tag{12.40}$$

    which approximates to $r_{bb'} + r_{b'e}$ for large values of $r_{b'c}$.

2. *Reverse voltage transfer ratio, h_{re}*: this is the ratio v_1/v_2 with the input open-circuited:

    $$h_{re} = \left.\frac{v_1}{v_2}\right|_{i_1=0} = \left.\frac{v_{be}}{v_{ce}}\right|_{i_b=0} \tag{12.41}$$

    This parameter serves the same purpose as the horizontal elements $r_{b'c}$ and $C_{b'c}$ in the hybrid-π model, in causing the input side to be affected by what is happening on the output side. Referring to Figure 12.8, the

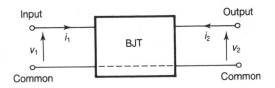

Figure 12.13 *h*-parameter model

voltage developed at the input due to a voltage v_{ce} applied to the output is:

$$v_{be} = \frac{v_{ce}r_{b'e}}{r_{b'e} + r_{b'c}} \tag{12.42}$$

Hence:

$$h_{re} = \frac{r_{b'e}}{r_{b'e} + r_{b'c}} \simeq \frac{r_{b'e}}{r_{b'c}} \tag{12.43}$$

This parameter has a small value in a good BJT, in which $r_{b'e} \ll r_{b'c}$.

3. *Forward current transfer ratio, h_{fe}:* this is the ratio of i_2/i_1 with the output short-circuited:

$$h_{fe} = \left.\frac{i_2}{i_1}\right|_{v_2=0} = \left.\frac{i_c}{i_b}\right|_{v_{ce}=0} \tag{12.44}$$

This is identical to the definition of the low frequency β in equation (12.11), so:

$$h_{fe} = \beta \tag{12.45}$$

The two terms are used interchangeably, although they arise in different models.

4. *Output conductance, h_{oe}:* this is the ratio of output current and voltage with the input open circuited:

$$h_{oe} = \left.\frac{i_2}{v_2}\right|_{i_1=0} = \left.\frac{i_c}{v_{ce}}\right|_{i_b=0} \tag{12.46}$$

i_c, v_{ce} and $v_{b'e}$ are related in the hybrid-π model by equation (12.20). If $v_{b'e}$ is substituted using equation (12.42) we obtain:

$$h_{oe} = \frac{g_m r_{b'e}}{r_{b'c}} + \frac{1}{r_{ce}} \tag{12.47}$$

The term $1/r_{ce}$ is the obvious output conductance, but the other term must be included if the effects of $r_{b'c}$ are significant.

The *h*-parameters are so named because they are hybrid quantities, being a mixture of dimensionless ratios and quantities with units. The close correspondence in name and structure with the hybrid-π model is a cause for confusion. The π structure is the only helpful part of the names, suggesting the form of that circuit model. Note that a simplified *h*-parameter circuit which neglects both h_{re} and h_{oe} is identical with the simplest form of the hybrid-π model.

Values of *h*-parameters are easily measured and are often quoted, together with their variation with temperature, so some familiarity with the quantities is necessary to the circuit designer. Those working with integrated devices might do better to concentrate on the hybrid-π version, which is a more accurate representation of the physical nature of the device.

EXAMPLE

■ Find *h*-parameter values for a typical BJT in which $\beta = 100$, $r_{bb'} = 20\ \Omega$, $r_{b'e} = 1\ k\Omega$, $r_{b'c} = 10\ M\Omega$ and $r_{ce} = 100\ k\Omega$.

☐ Equation (12.40) gives h_{ie}, the input resistance:

$$h_{ie} = 20 + \frac{10^3 \times 10^7}{10^3 + 10^7}$$

$$= 1{,}019.9\ \Omega \quad (\simeq r_{b'b} + r_{b'e})$$

Note that $r_{b'c}$ is large enough to be omitted. Equation (12.43) gives the reverse voltage transfer ratio:

$$h_{re} = \frac{10^3}{10^3 + 10^7} \simeq 10^{-4}$$

The small value shows that the output conditions do not affect the input very strongly. The common emitter current gain is identical with the forward current transfer ratio, so:

$$h_{fe} = \beta = 100$$

Using equation (12.47) with equation (12.36) for g_m, the output conductance is:

$$h_{oe} = \frac{100}{10^7} + \frac{1}{10^6}$$

$$= 1.1 \times 10^{-5}\ S$$

12.3 Switching

The majority of integrated devices are used for digital electronics, in which the signals take either a high or a low value, corresponding to the two digits of binary arithmetic. Any two voltage or current levels could be used to denote 0 and 1, for instance 4.3 V = 0 and 4.5 V = 1 (Figure 12.14). In this case the BJT would make small transitions within its active region, and the small-signal ac models developed above could be used. However, it would only need a noise component of 0.2 V amplitude to make a nonsense of the digital data. Therefore it is usual to choose two well separated voltage levels to provide noise immunity.

The levels used are the cut off and saturation conditions, as in Figure 12.4b. The BJT is driven through the active region rapidly in large switching transitions, which invalidates the small-signal assumption in the hybrid-π and h-parameter models. The questions to be settled are:

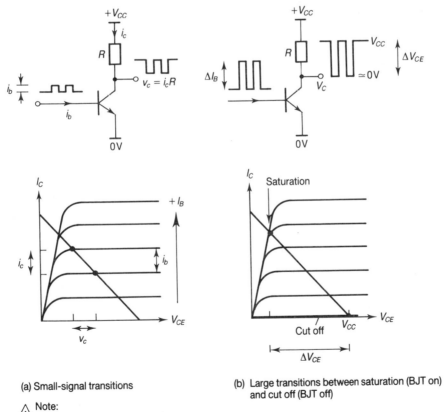

(a) Small-signal transitions

(b) Large transitions between saturation (BJT on) and cut off (BJT off)

 Note:
Biasing components are not shown.

Figure 12.14 Switching between cut off and saturation

1. How quickly can a BJT switch between cut off and saturation?
2. What can the device designer or the circuit designer do to minimize the switching time?

While a BJT is in the process of switching its state is undefined as far as digital circuitry is concerned. If it takes T seconds to change state, the maximum frequency of information transfer is $1/T$ Hertz. Fast data processing therefore requires a short switching time.

12.3.1 Turn on

Consider a transistor connected in common emitter mode which is biased to cut off, such that $I_B = 0$ and the excess injected charge in the base $\hat{Q}_B = 0$. The polarity of this charge does not matter because there is a balancing charge supplied through the base current (section 11.3). This analysis therefore applies equally to *npn* and *pnp* transistors. For a fixed value of base current $\hat{Q}_B$ and I_B are related by the charge control condition:

$$I_B = \frac{\hat{Q}_B}{\tau_B} \text{ A} \tag{12.48}$$

τ_B is the minority carrier lifetime in the base. If I_B varies with time, some part of the current goes to alter the excess charge in the base so that equation (12.48) becomes:

$$I_B = \frac{\hat{Q}_B}{\tau_B} + \frac{d\hat{Q}_B}{dt} \text{ A} \tag{12.49}$$

Now suppose that the base current takes a sudden jump up from zero to a value I_{B1} at time $t = 0$. The excess injected charge builds up in response, after an initial delay t_d while the charge stored in the emitter–base depletion region adjusts to the new bias condition. The switching is usually dominated by t_s, the switching time during which the excess injected charge builds up to the saturation value. The variables in equation (12.49) can be separated to give:

$$\int_0^{\hat{Q}_B} \frac{\tau_B d\hat{Q}_B}{\tau_B I_{B1} - \hat{Q}_B} = \int_0^t dt \quad (t > 0) \tag{12.50}$$

which has a solution:

$$t = \tau_B \ln \frac{\tau_B I_{B1}}{\tau_B I_B - \hat{Q}_B} \text{ s} \tag{12.51}$$

Making use of the relationship between excess charge in the base and collector current, $I_C = \hat{Q}_B / \tau_t$, and equation (11.27) for β, the time t_s to reach the saturation value I_{Csat} is:

$$t_s = \tau_B \ln \frac{\beta I_{B1}}{\beta I_{B1} - I_{Csat}} \text{ s} \qquad (12.52)$$

Note that this relation assumes that the value of the base current I_{B1} is sufficient to drive the transistor into saturation. In the limit that $\beta I_{B1} = I_{Csat}$ the time to reach saturation becomes infinite, like the charging of a capacitor to its full dc voltage.

Equation (12.52) suggests three ways to minimize the turn on time, two of which are properties of the BJT, while one is determined by the external circuit:

1. *Make τ_B small.* The minority carrier lifetime in the base is the character-istic timescale for changes in $\hat{Q}_B$ and I_C. This is seen by rewriting equation (12.51):

    $$\hat{Q}_B = I_{B1}\tau_B(1 - \exp -t/\tau_B) \qquad (12.53)$$

    where the exponential decay period is keyed to τ_B.

2. *Make β large.* In the limit of $\beta I_{B1} \gg I_{Csat}$ the switching time tends to zero. Note that there is a compromise between these two conditions, because $\beta = \tau_B/\tau_t$. Decreasing τ_B will cause β to fall unless the transit time τ_t across the base is also reduced.

3. *Make I_{B1} large.* The circuit designer can do one thing to speed up the switch on, by driving the transistor hard on with a large base current. The ratio of base current to that which just causes saturation is called the base overdrive. When a large step in I_B is applied the collector current rises as:

    $$I_C = \frac{I_{B1}\tau_B}{\tau_t} (1 - \exp -t/\tau_B) \qquad (12.54)$$

    until it reaches I_{Csat}. Figure 12.15 shows how a large value of I_{B1} reduces t_s by catching the exponential curve early on.

Having reached saturation the collector–emitter voltage falls to its lowest level, about 0.2 V, as the collector–base diode becomes forward biased. The collector voltage alters very little with further changes in base current, so the collector current is also fixed at the value determined by the external load resistance. This corresponds to the coalescence of the lines for different values of base current at low V_{CE} on the common emitter I_C/V_{CE} characteristic (Figure 12.3b).

Although the collector current stops varying when it reaches the saturation value increases in base current do cause further changes in injected charge within the base. Figure 12.16 shows the build up of $\hat{Q}_B$ as a BJT is switched into saturation with base overdrive. At saturation the collector also emits minority carriers into the base, instead of collecting all such carriers at the depletion

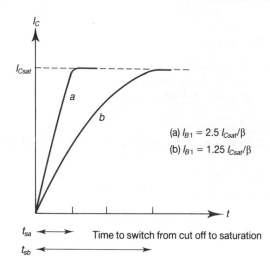

Figure 12.15 A larger base overdrive makes I_C reach I_{Csat} more quickly

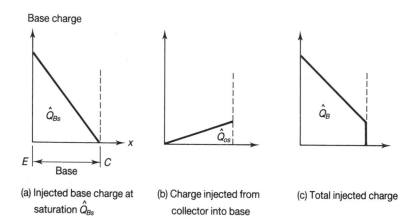

Figure 12.16 Build up of Q_B as a BJT is switched into saturation with base overdrive

region edge. The charge distribution is then the sum of two short diode triangular profiles, one for each forward biased junction. This oversaturation charge $\hat{Q}_{os}$ has a different lifetime τ_{os} from that injected only from the emitter, typically about half τ_B. In saturation equation (12.49) becomes:

$$I_B = \frac{\hat{Q}_{Bs}}{\tau_b} + \frac{\hat{Q}_{os}}{\tau_{os}} + \frac{d\hat{Q}_{os}}{dt} \tag{12.55}$$

Note that only the oversaturation charge varies once $\hat{Q}_B$ achieves its saturation value Q_{Bs}. This has a solution:

$$\hat{Q}_{os} = \tau_{os}\left(I_{B1} - \frac{\hat{Q}_{Bs}}{\tau_B}\right)(1 - \exp -t/\tau_{os}) \qquad (12.56)$$

where $t = 0$ when the BJT enters saturation, not when it switches from cut off. The degree of oversaturation affects the switch off time of the device, as discussed below.

12.3.2 Turn off

If the base current is quickly reduced at $t = 0$ to a lower value I_{B2} from a fully on value the sequence described for turn on is reversed. The fall through oversaturation is again given by equation (12.55), with different limits. At $t = 0$ the oversaturation charge $\hat{Q}_{os}$ is not zero, but takes some initial value $\hat{Q}_{os1}$, which may be as high as $\tau_{os}(I_{B1} - \hat{Q}_{Bs}/\tau_B)$ from equation (12.56). The solution in the turn off case contains the term $\hat{Q}_{os} \exp -t/\tau_{os}$ which shows that the decay through oversaturation is keyed to the time constant τ_{os}.

Until $\hat{Q}_{os}$ falls to zero the collector current is pinned to the saturation value and the transistor cannot begin to switch off. This period is called the storage delay time, t_{sd}, as for the junction diode. Note that although the circuit effects are similar in BJTs and diodes – they both stay on during the storage delay time – the physical causes are slightly different.

The transistor enters the active region as soon as the oversaturation charge has been dissipated. Equation (12.49) then becomes valid, with the initial condition altered to $\hat{Q}_B = \hat{Q}_{Bs}$ at $t = 0$, measuring time from the transition point. The solution is:

$$\hat{Q}_B = \tau_B I_{B2}(1 - \exp -t/\tau_B) + \hat{Q}_{Bs} \exp -t/\tau_B \qquad (12.57)$$

If the switch off base current I_{B2} is zero this gives an exponential decay of Q_B and hence of collector current, with a time constant τ_B. Once again the circuit designer can speed things up considerably by making I_{B2} negative. This sucks out the excess charge, rather than letting it recombine gradually. Equation (12.57) is only valid for $0 < \hat{Q}_B < \hat{Q}_{Bs}$, because the transistor cuts off once the excess base charge has been removed.

The base and collector current waveforms during switching are displayed in Figure 12.17. A single figure of merit is frequently given to BJTs, rather than defining a series of time constants. The quantity usually quoted is the rise time t_r or fall time of the collector current between 10% and 90% of the total change in current between cut off and saturation. This excludes the delays associated with depletion region charging and oversaturation charge dissipation, but measures the main $\exp -t/\tau_B$ term:

$$t_r = 2.2\tau_B \qquad (12.58)$$

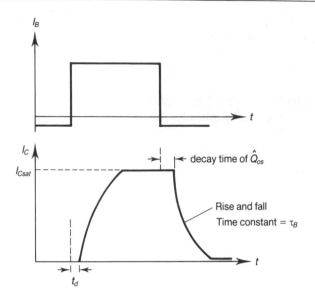

Figure 12.17 BJT switching response

A typical value for t_r is 30 ns for a small-signal transistor. The discussion above shows that the quoted value should not be used blindly, because t_r is a function of the base current waveform.

EXAMPLE

■ A BJT in a common emitter circuit has a collector saturation current of 200 mA, a base transit time of 0.5 microseconds and a base minority carrier lifetime of 30 microseconds. Find the switch on time t_s for a base current step increase from zero to (1) 4 mA; (2) 10 mA.

☐ Equations (12.52) or (12.53) can be used, with:

$$\beta = \frac{\tau_B}{\tau_t} = \frac{30}{0.5} = 60$$

The minimum base current to achieve saturation is $200/60 = 3.3$ mA, so both 4 mA and 10 mA are into overdrive.

1. 4 mA

$$t_s = 0.5 \ln \frac{60 \times 4}{60 \times 4 - 200}$$

$$= 0.9 \, \mu s$$

2. 10 mA

$$5_s = 0.5 \ln \frac{60 \times 10}{60 \times 10 - 200}$$

$$= 0.2 \, \mu s$$

Increasing the overdrive clearly decreases the switching time. Note that the expression was evaluated in terms of currents in mA. This is permissible only because the currents appear as ratios, so that any units can be used provided the same units are used throughout.

12.4 Real bipolar devices

This section contains descriptions of the commonly available bipolar junction devices, including power variants and some small-scale integration (SSI) packages which hold BJTs in standard circuit configurations.

12.4.1 Low power BJTs

Most of the preceding discussion on BJTs has referred to low power devices, typically with less than one watt power dissipation capacity. These are sometimes confusingly labelled small-signal transistors, although they can handle both digital switching and true small signals. The parameters given here are typical of a range of discrete (i.e. separately packaged) BJTs:

1. *Maximum power dissipation*: 0.3–1 W. This is usually determined by the thermal resistance of the package. Plastic packages are good at the lowest powers, then metal packages, then packages which can be bolted directly to a heatsink for the highest powers. The object is always to keep the junction temperature below about 150 °C for silicon.

2. *Common emitter current gain* β: 80–800. It is very difficult to make a transistor with a precise value of β so the specification is usually broad. Circuit designs should not rely on a particular value, but should use the fact that I_C/I_B is normally large.

3. *Maximum collector current* I_C: 100–800 mA. This is specified separately from the power to avoid high level injection effects such as current crowding.

4.   *Maximum collector–emitter voltage* V_{CEO}: 10–70 V before breakdown.

5.   *Maximum collector–base voltage*: 30–80 V before breakdown. Special high voltage devices can extend V_{CEO} and V_{CBO} to a few hundred volts, sacrificing gain and bandwidth.

6.   *Gain–bandwidth product* f_T: 1–400 MHz, typically 200 MHz. This applies to small sinusoidal signals, not to switching between cut off and saturation.

12.4.2 High power BJTs

A wide range of medium and high power BJTs are available, with different applications: low voltage and high current; high voltage and low current; or some compromise. These devices nearly all require bolting to a chassis or heatsink. It is difficult to obtain high power, high gain and wide bandwidth in one device. Maximum ratings for a typical compromise device are:

1.   Power 120 W.
2.   Current 15 A.
3. β: 20–70.
4. V_{CEO}: 60 V.
5. V_{CBO}: 100 V.
6. f_T: 1 MHz.

12.4.3 Darlington pairs

Two BJTs connected as in Figure 12.18 are known as a Darlington pair. The resulting three-terminal unit behaves like a single BJT with a common emitter

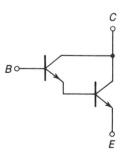

Figure 12.18 Darlington pair configuration

current gain β_{Dar} equal to the product of the individual transistor betas:

$$\beta_{Dar} = \beta_1\beta_2 \tag{12.58}$$

Both low and high power transistors are available in a single Darlington package, with β_{Dar} up to several hundred. The base–emitter voltage is the sum of two forward biased diode drops and V_{CEsat} is at least one diode drop. Switching speed is sacrificed in favour of gain in the Darlington connection. It should not be confused with superbeta transistors, which are single BJTs designed for extremely high gain.

12.4.4 Complementary pairs

Analogue circuits running between positive and negative voltage rails often demand symmetrical behaviour from the positive- and negative-driven halves. *Npn* transistors in common emitter mode require V_{CE} positive, while *pnp* transistors use a negative V_{CE}. *Pnp* and *npn* transistors are therefore offered as matched or complementary pairs, with similar I/V characteristics and matched temperature coefficients. Both low and high power pairs are available.

12.4.5 Four-layer devices

In this section we group together a class of devices having in common a four-layer *npnp* structure. Both two- and three- terminal versions are made, so the discussion overlaps diode and bipolar junction transistor topics. The other universal feature of these devices is their aim of mimicking the action of a switch (Figure 12.19). An ideal switch passes no current regardless of the voltage across it, when open and passes unlimited current with no voltage drop when closed. Mechanical contacts such as relays approximate to this, with hold off voltages of the order of 2 kV and a contact resistance when closed of about 0.2 ohms. They are also slow, taking tens of milliseconds to close or open, and have a restricted life due to wear of the contacts.

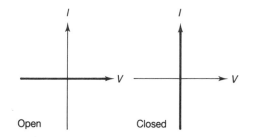

Figure 12.19 *I/V* characteristics of an ideal switch

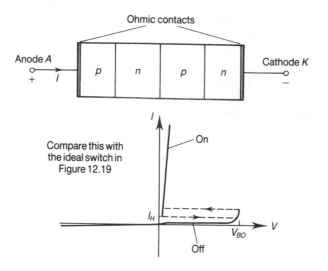

Figure 12.20 Four-layer *pnpn* diode

The simplest four-layer structure has ohmic contacts at both ends (Figure 12.20). The *p*-type end is called the anode and the *n*-type end the cathode. When the anode is taken positive the outer junctions J_1 and J_3 are forward biased, while the inner junction J_2 is reverse biased. Only a leakage current flows up to a voltage V_{BO}, the *breakover voltage*, at which point the device changes state abruptly. The current rises sharply and the voltage drop collapses to a low value, typically 0.7 V. If the current is reduced below the *holding current* I_H, the device reverts to its high impedance blocking state.

The conventional explanation for this behaviour is to regard the structure as two bipolar transistors, as in Figure 12.21. Q_1 is *pnp* and Q_2 is *npn*, with collector leakage currents I_{C01} and I_{C02} respectively. Using equation (11.12) for each transistor,

$$I_{C1} = -\alpha_1 I + I_{C01}$$

$$I_{C2} = \alpha_2 I + I_{C02}$$

and the current summation for Q_1 is:

$$I + I_{C1} - I_{C2} = 0$$

Hence:

$$I = \frac{I_{C0}}{1 - (\alpha_1 + \alpha_2)}$$

where $I_{C0} = I_{C02} - I_{C01}$. The values of α_1 and α_2 depend on the voltages across

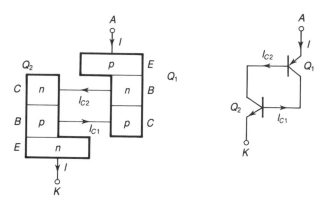

Figure 12.21 Two-BJT model of the *pnpn* diode

Q_1 and Q_2, particularly when these are high enough to multiply the number of charges by the avalanche breakdown mechanism. The current I rises to the limit set by the external circuitry when $(\alpha_1 + \alpha_2) = 1$.

An alternative explanation can be given in terms of the band diagram of the *pnpn* structure, Figure 12.22. Assuming three symmetrical junctions the zero bias diagram in Figure 12.22a shows a constant Fermi level. For forward bias up to V_{BO} the band diagram takes the form shown in Figure (12.22b). The two outer junctions are very slightly forward biased and the central junction takes up most of the applied potential in reverse bias.

When the breakover voltage is reached avalanche multiplication creates a large number of electrons and holes in the central high field region. These are swept apart by the field into the neighbouring *p*- and *n*-type regions. The sudden introduction of these charges cannot be balanced instantly by injection across the forward biased junctions, so the *n*-type region goes negative and the *p*-type side goes positive. The band diagram (Figure 12.22c) reveals that this reduces the reverse bias on the central junction, and *can even take it into forward bias*.

The same change takes the outer junctions strongly forward biased. They respond by injecting large numbers of carriers towards the central junction. A new steady state condition is set up with *all three* junctions forward biased, but with two of the voltage drops in opposition. Note that it is not possible to view this as three *separate* forward bias *pn* diodes. A steady state requires a collective behaviour, as in Figure 12.22c, where both inner regions act as the base of a bipolar transistor. Carriers injected from the outer junctions diffuse across the next region to be collected as minority carriers on the other side.

This double transistor action serves to continue the flow of charges in the sense begun by avalanche multiplication at breakover. If the current is reduced so that the forward bias injection component in the central junction exceeds the

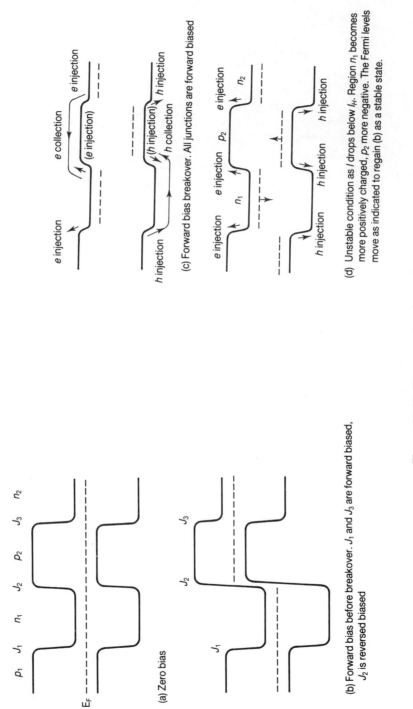

Figure 12.22 The *pnpn* four-layer diode

(a) Zero bias

(b) Forward bias before breakover. J_1 and J_3 are forward biased, J_2 is reversed biased

(c) Forward bias breakover. All junctions are forward biased

(d) Unstable condition as I drops below I_H. Region n_1 becomes more positively charged, p_2 more negative. The Fermi levels move as indicated to regain (b) as a stable state.

collected component, the unstable condition in Figure 12.22d is set up, which rapidly takes the central junction into reverse bias again.

The addition of a third electrode to the inner p-type region allows the breakover voltage to be controlled externally. When the gate electrode is biased positive with respect to the adjacent n-type cathode, this junction is forward biased. Part of the current injected into the p-type region is collected by the central junction, where it reinforces the leakage current in the hold off condition and hence brings about an earlier transition to the conducting state. This configuration is called a *thyristor* or a semiconductor controlled rectifier (SCR). Its circuit symbol and I/V characteristic are shown in Figure 12.23.

The thyristor is widely used to control a few hundred watts of power at a few hundred volts, such as in light dimmer switches in homes. The power is

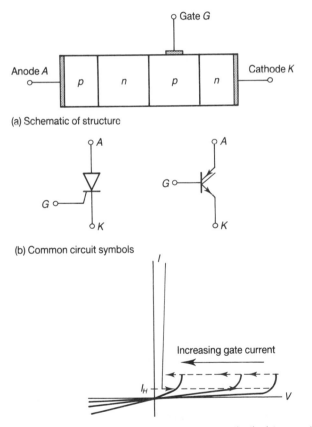

(a) Schematic of structure

(b) Common circuit symbols

(c) I/V characteristics. A pulse of gate current turns the thyristor on until I falls below I_H

Figure 12.23 The SCR or thyristor

controlled by holding the device off for part of the alternating cycle, then firing it on until the current drops towards zero. The resulting harmonics will doubtless cause headaches to the power supply authorities as more and more of their loads are controlled in this way.

A single thyristor can only be fired once every full cycle, and lies fallow during the reverse half cycle. A variant which overcomes this is called the *triac*. This consists of two thyristors combined in a single device, to make use of both half cycles. Another variant called the gate turn off thyristor (GTO thyristor) possesses a gate which can interrupt the on condition by the application of a current pulse. This makes it unnecessary to wait for an ac current zero to turn off the device.

Summary

Main symbols introduced

Symbol	Unit	Description
f_β	Hz	frequency for 45° phase shift in gain
f_T	Hz	transition frequency, or gain–bandwidth product
g_m	Ω^{-1}	mutual conductance, hybrid-π model
h_{fe}		h-parameter, forward current transfer ratio
h_{ie}	Ω	h-parameter, input resistance
h_{oe}	Ω^{-1}	h-parameter, output conductance
h_{re}		h-parameter, reverse voltage transfer ratio
i	A	small-signal alternating current
I_H	A	SCR holding current
$\hat{Q}_{Bs}$	C	excess base charge in saturation
$\hat{Q}_{os}$	C	excess base charge over $\hat{Q}_{Bs}$ in oversaturation
$r_{bb'}$	Ω	base-spreading resistance, hybrid-π model
$r_{b'c}$	Ω	collector–base resistance, hybrid-π model
r_{be}	Ω	input resistance, hybrid-π model
r_{ce}	Ω	output resistance, hybrid-π model
t_d	s	switching delay time before I_C responds from cut off
t_r	s	current risetime between 10% and 90% in a switching transition
t_s	s	time for I_C to reach saturation in switching
V_{BO}	V	SCR breakover voltage
β_{Dar}		current gain of Darlington pair
τ_{os}	s	lifetime of excess oversaturation charge

Main equations

$$\alpha = -\left.\frac{\partial I_C}{\partial I_E}\right|_{V_{CB}=\text{constant}} = -\frac{i_c}{i_e} \tag{12.3, 12.10}$$

$$\beta = \left.\frac{\partial I_C}{\partial I_B}\right|_{V_{CE}} = \frac{i_c}{i_b} \tag{12.4, 12.11}$$

$$g_m \simeq \frac{qI_E}{kT} \tag{12.18}$$

$$\beta = g_m r_{be} = h_{fe} \quad (f < f_\beta) \tag{12.19, 12.45}$$

$$|\beta f| = g_m r_{b'e} f_\beta = f_T \quad (f \gg f_\beta) \tag{12.34}$$

$$I_B = \frac{\hat{Q}_B}{\tau_B} + \frac{\mathrm{d}\hat{Q}_B}{\mathrm{d}t} \quad \text{(presaturation)} \tag{12.49}$$

$$I_B = \frac{\hat{Q}_{Bs}}{\tau_B} + \frac{\hat{Q}_{os}}{\tau_{os}} + \frac{\mathrm{d}\hat{Q}_{os}}{\mathrm{d}t} \quad \text{(oversaturation)} \tag{12.54}$$

Exercises

1. Fill in the blanks in this table.

Type	Bias condition	V_{BE}	V_{CB}	V_{CE}
npn	active	0.7	5.0	5.7
npn		0.6		10.0
pnp		−0.6	−0.3	
	saturation	0.7		0.2
pnp			−6.0	−5.4

2. The input resistance of a BJT is 50 Ω at a junction temperature of 70 °C with an emitter current of 40 mA. Find the common emitter current gain at 100 °C with a current of 50 mA.

3. The common emitter hybrid-π parameters of a BJT are: $r_{bb'} = 15\ \Omega$, $r_{b'e} = 75\ \Omega$, $\beta = 100$, $r_{ce} = 150\ \mathrm{k}\Omega$. A 2 V peak to peak low frequency sinusoidal signal is applied to the base via a 1 kΩ series resistance. Find the peak to peak value of the signal across a 100 Ω load resistance. Neglect $r_{b'c}$.

4. Injected minority carriers cross a 10 μm wide base at the saturation velocity, $10^5\ \mathrm{m\,s^{-1}}$. Find the base–emitter capacitance at an emitter current of 80 mA, and the gain–bandwidth product for a junction temperature of 100 °C.

5.   A BJT has a transition frequency of 1 MHz. Estimate the base–emitter capacitance when running at $I_E = 0.5$ A if the low frequency value of β is 100. What is the maximum frequency that capacitances can be neglected?

6.   Estimate the low frequency hybrid-π parameters of a BJT with h-parameters: $h_{ie} = 1 \text{ k}\Omega$, $h_{re} = 10^{-4}$, $h_{fe} = 90$, $h_{oe} = 10^{-5}$ S. (Neglect $r_{bb'}$.)

7.   Find the time t_s to switch between cut off and saturation with a base overdrive of 3, if the minority carrier lifetime in the base is 20 μs.

8.   Verify that equation (12.55) follows from equation (12.54). Find the solution to equation (12.54) for the switching off case, where the oversaturation charge is falling from a value $\hat{Q}_{os1}$ at time $t = 0$. What could be done to minimize this component of the switching time?

9.   Schottky-clamped BJTs possess a Schottky barrier diode in parallel with the collector–base diode, with the same polarity. How will this affect the switch on time and switch off time?

10.   A heterojunction transistor has an emitter made from a wider band gap material, such as AlGaAs, while the GaAs base and collector have a smaller band gap. Sketch an equilibrium band diagram for such a device and suggest how the emitter injection efficiency is altered compared to an ordinary homojunction BJT.

13 The junction field effect transistor

13.1 Introduction

The junction field effect transistor or JFET is one example of a class of devices called field effect transistors or FETs. Like the bipolar transistor the purpose of an FET is either to amplify a small signal or to act as a digital switch. The FET has three terminals: source, drain, and gate. In normal use the load current flows between source and drain, while the input is applied to the gate.

There is one important difference between bipolar junction transistors and FETs:

1. BJTs are controlled by the *current* flowing into the base.
2. FETs are controlled by the *voltage* applied to the gate.

The name field effect transistor refers to the controlling influence of the electric field within the gate area of the device. FETs have a high input impedance, normally well in excess of $10\,\text{M}\Omega$, which accords with the notion of voltage- rather than current-control.

The FET family can be divided into three groups, as in Figure 13.1,

275

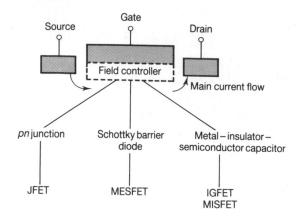

Figure 13.1 FET family

according to the cause of the high input impedance. These are:

1. Reverse biased *pn* junction: the junction FET or JFET.
2. Reverse biased Schottky barrier diode: the metal–semiconductor FET or MESFET.
3. An insulating layer: the metal–insulator semiconductor FET or MISFET.

The JFET and MESFET will be described in this chapter and the MISFET in the following two chapters.

The internal structure of a JFET is shown in Figure 13.2. The channel

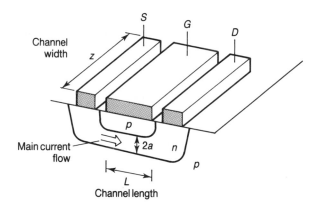

Note:
The channel width, z, height, $2a$, and length, L, are shown.

Figure 13.2 Internal structure of an *n*-channel JFET

width z is usually much greater than the length L, in contrast to the usual use of the words width and length. The width determines a JFET's ability to handle current, while its length is critical to its I/V characteristic. This example is called an n-channel device because the current path between source and drain is entirely within n-type material. JFETs are sometimes called unipolar transistors because only majority carriers are responsible for current flow, in contrast to bipolar devices where both majority and minority carriers are used. With no minority carrier populations to be dissipated, JFETs can have switching times as low as 0.1 ns for GaAs MESFETs and 0.3 ns for silicon JFETs.

Circuit symbols for n-channel and p-channel JFETs are shown in Figure 13.3. The arrow is in the same sense as the gate–channel pn diode.

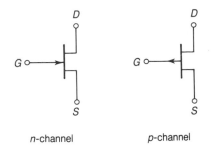

n-channel p-channel

Note:
The arrow is in the same sense as the gate – channel diode.

Figure 13.3 Circuit symbols for JFETs and MESFETs

13.2 Before pinch off

Consider an n-channel JFET with dimensions as shown in Figure 13.2. It is assumed that there is a p^+ layer also at gate potential under the channel in order to make the geometry symmetrical. The gate is doped p^+, so that the gate–channel depletion region resides mostly within the channel. It is the variation in channel width with applied gate voltage which alters the current flow between source and drain.

First suppose the source and drain are connected, so that $V_{DS} = 0$, and a negative gate–source voltage V_{GS} is applied (Figure 13.4a). The gate–channel and substrate–channel diodes are reverse biased in this n-channel JFET, so a depletion region expands into the more lightly doped channel. The depletion region has the same width everywhere, because the same reverse bias exists at all points along the channel in the absence of current flow.

Now suppose the source is grounded and the drain is made positive by V_{DS} volts. A negative potential V_{GS} (<0) is applied to the gate, so that the gate–channel diode is still reverse biased. At the source end the voltage V_{GCH} across this gate–channel diode is:

$$V_{GCH}(0) = \text{gate potential} - \text{source potential}$$

$$= V_{GS} \; (<0) \qquad\qquad (13.1)$$

while at the drain end the potential is:

$$V_{GCH}(L) = \text{gate potential} - \text{drain potential}$$

$$= V_{GS} - V_{DS} \qquad\qquad (13.2)$$

Clearly there is a greater reverse bias at the drain end, and hence a thicker depletion region there (Figure 13.4b). This has consequences for the current flowing along the channel, because it is being confined to flow in a smaller cross-sectional area as it progresses from source to drain. The band diagram of a vertical section through the channel confirms this (Figure 13.5): electrons face a large potential barrier on either side. The barriers not only confine electrons to the n-type channel, but also deny some of the channel region to them.

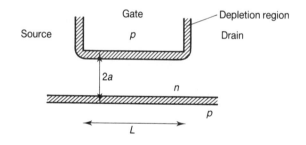

(a) $V_{DS} = 0$, $V_{GS} < 0$ No channel current and a constant thickness of depletion region

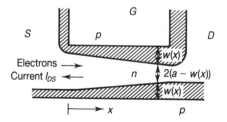

(b) $V_{DS} > 0$, $V_{GS} < 0$ A source–drain current flows and the depletion region thickness varies along the channel. The open channel height is $2(a - w(x))$ at position x

Figure 13.4 JFET channel geometry

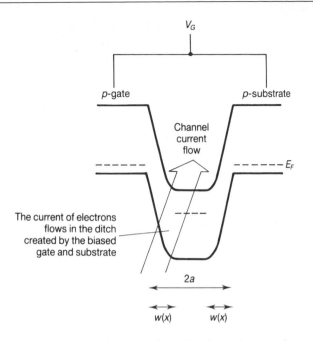

Figure 13.5 Band diagram of a vertical section through an *n*-channel JFET

The resistance of the channel therefore varies along its length, because the open cross-sectional area varies. Worse, the amount of open channel depends on the voltage drop along the channel, which in turn depends on the resistance! We will go through the analysis which disentangles this next. It may be skipped without loss of continuity, but degree students should realize that hard pressed lecturers often dish it up as a question in examinations.

13.2.1 *I/V characteristic before pinch off*

The main assumptions in the calculation of the *I/V* characteristic are:

1. The channel length L is large compared to its width, $2a$. This is called the gradual channel approximation. Its main implication is that current only flows along the channel. Resistive losses due to current flowing vertically as the channel becomes narrower are neglected.
2. The gate potential is applied to the substrate as well.
3. Gate and substrate doping levels N_A are equal and uniform.
4. The channel doping N_D is uniform.
5. $N_A \gg N_D$, so that single-sided *pn* diode approximations can be used.

The depletion region width on each side of the channel (above and below) is $w(x)$, where x is the position along the channel, as shown in Figure 13.4b. The value of $w(x)$ depends on $V_{GCH}(x)$, the local value of the gate–channel potential difference:

$$w(x) = \left(\frac{2\epsilon_0\epsilon_r|V_{GCH}(x)|}{qN_D}\right)^{1/2} \text{ m} \tag{13.3}$$

The current flows through a cross-sectional area $A(x)$, which varies along the channel:

$$A(x) = z(2a - 2w(x)) \text{ m}^2 \tag{13.4}$$

where z is the channel width defined in Figure 13.2. Now consider the current I_{SD} which flows through a narrow slice of the channel (Figure 13.6). I_{SD} is the conventional current flowing in the $+x$ direction from source to drain. The conductivity of the n-type slice is:

$$\sigma = N_D q \mu_e \; \Omega^{-1} \tag{13.5}$$

and Ohm's law for the slice is:

$$I_{SD} = A(x)J_{SD} = A(x)\sigma\mathscr{E}(x) = -A(x)N_Dq\mu_e\frac{\mathrm{d}V_{CH}}{\mathrm{d}x} \tag{13.6}$$

Here $V_{CH}(x)$ is the local channel potential with respect to the grounded source. Hence $V_{CH}(0) = 0$ and $V_{CH}(L) = V_{DS}$. The sign conventions for current stated in chapter one have been used. The conventional current I_{DS} flowing from drain to source is simply given by:

$$I_{DS} = -I_{SD} \tag{13.7}$$

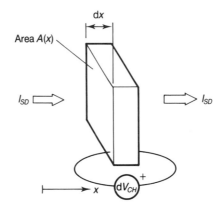

Figure 13.6 Current flow in a small slice of the channel

Substituting for I_{SD}, $A(x)$ and $w(x)$ gives:

$$I_{DS} = +2zqN_D\mu_e \left[a - \left(\frac{2\epsilon_0\epsilon_r(V_{CH}(x) - V_{GS})}{qN_D} \right)^{1/2} \right] \frac{dV_{CH}}{dx} \tag{13.8}$$

Now Kirchoff's current law implies that if I_{DS} flows in at one end of the channel, it flows out of the other end, *and* through all points in between. This means that I_{DS} does not vary with x, and we have an equation, albeit a messy one, which only has $V_{CH}(x)$ depending on x. To solve it we integrate both sides with respect to x using the ends of the channel as limits of integration:

$$\int_0^L I_{DS}\, dx = +2zqN_D\mu_e \int_0^{V_{DS}} a - \left(\frac{2\epsilon_0\epsilon_r}{qN_D} \right)^{1/2} (V_{CH} - V_{GS})^{1/2}\, dV_{CH} \tag{13.9}$$

or

$$\frac{LI_{DS}}{2zqN_D\mu_e} = \left[+aV_{CH} - \left(\frac{2\epsilon_0\epsilon_r}{qN_D} \right)^{1/2} \frac{2}{3}(V_{CH} - V_{GS})^{3/2} \right]_0^{V_{DS}}$$

$$= aV_{DS} - \frac{2}{3}\left(\frac{2\epsilon_0\epsilon_r}{qN_D} \right)^{1/2} [(V_{DS} - V_{GS})^{3/2} - (-V_{GS})^{3/2}] \tag{13.10}$$

This can be tidied up a little by defining:

$$G_o = \frac{2zqaN_D\mu e}{L}\ \Omega^{-1} \tag{13.11}$$

as the channel conductance at zero gate voltage and very low drain–source voltage. In this limit the JFET channel behaves like a plain linear resistance $1/G_o$ ohms. Another quantity of importance is the *pinch off voltage*, V_p, which is the gate to channel potential difference necessary to bring the two depletion regions together. This is found by putting $w = a$ in equation (13.3):

$$V_p = \frac{-qa^2N_D}{2\epsilon_0\epsilon_r} \tag{13.12}$$

Using these quantities the I/V characteristic before pinch off becomes:

$$I_{DS} = G_o \left[V_{DS} + \frac{2V_p}{3} \left\{ \left(\frac{V_{GS} - V_{DS}}{V_p} \right)^{3/2} - \left(\frac{V_{GS}}{V_p} \right)^{3/2} \right\} \right] \tag{13.13}$$

The equation is plotted in Figure 13.7 for various values of V_{GS}. It is only valid below pinch off because we have assumed Ohm's law conduction through the open part of the channel.

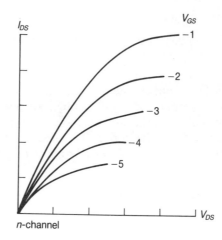

Figure 13.7 n-channel JFET I_{DS}/V_{DS} characteristics before pinch off

13.3 After pinch off

If the gate of an n-channel JFET is made steadily more negative, or the drain–source voltage is increased, the depletion region expands into the channel. There comes a point at which the depletion regions from the gate and the substrate meet in the centre of the channel (Figure 13.8). This is called pinch off, and occurs first at the drain end, where the gate–channel potential is largest:

$$V_{GCH} \text{ (drain end)} = V_{GS} - V_{DS} \qquad (13.14)$$

Remember V_{GS} is negative and V_{DS} is positive in an n-channel JFET, so $V_{GS} - V_{DS}$ is the most negative value of V_{GCH}. Pinch off occurs when:

$$V_{GS} - V_{DS} = V_p \qquad (13.15)$$

where the pinch off voltage V_p is defined in equation (13.12).

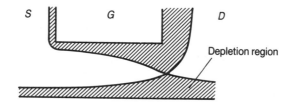

Figure 13.8 n-channel JFET at pinch off

EXAMPLE

■ Find the full channel resistance and pinch off voltage for a JFET with channel length, height and width of 25 μm, 2 μm and 200 μm. The channel is doped n-type with 10^{22} m^{-3} donors and is silicon. If V_{GS} is -3 V, what is the value of V_{DS} at pinch off?

☐ The full channel conductance is given by equation (13.11):

$$G_o = \frac{2 \times 200 \times 10^{-6} \times 1.6 \times 10^{-19} \times 10^{-6} \times 10^{22} \times 0.135}{25 \times 10^{-6}}$$

$$= 3.46 \times 10^{-3} \, \Omega$$

so the full channel resistance is $1/G_o$ or 289 Ω.

The pinch off voltage is given by equation (13.12):

$$V_p = \frac{-1.6 \times 10^{-19} \times (10^{-6})^2 \times 10^{22}}{2 \times 8.85 \times 10^{-12} \times 11.9}$$

$$= -7.6 \, \text{V}$$

Thus a gate–source voltage of -7.6 V will pinch off the channel for zero drain–source voltage. Alternatively, a drain–source voltage of $+7.6$ V will cause pinch off at the drain end if the gate and source are grounded. These are extremes of the general condition for pinch off, stated in equation (13.15), which causes the bias of the gate–channel diode to be V_p at the drain end. Hence if $V_{GS} = -3$ V, the value of V_{DS} at pinch off is:

$$V_{DS} = V_{GS} - V_p$$

$$= -3 + 7.6 \, \text{V}$$

$$= 4.6 \, \text{V}$$

Now what happens? At first sight the drain–source current I_{DS} might fall rapidly to zero as the conducting region is choked off. However, the band diagram along the channel reveals that this is not the case. Just before pinch off (Figure 13.9a) the diagram shows the Fermi level and band edges parallel and sloping down from source to drain. The slope increases towards the drain, indicating an increase in electric field there. The sloping Fermi level means that we are out of equilibrium, and a current flows: electrons roll down to the drain.

After pinch off the diagram resembles Figure 13.9b. A region of much greater slope now exists, with an upwards curvature indicating positive charge. However, no barrier has been set up by this charged region. Indeed, electrons entering it are whisked towards the drain at saturation velocity (see Figure 2.6).

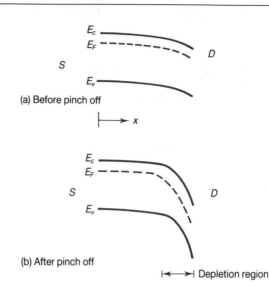

Figure 13.9 *n*-channel JFET band diagram along the channel

The net effect is to move the drain a little closer to the source, and to compress the channel shape just before pinch off into a slightly smaller length. Therefore:

The drain–source current after pinch off does not vary much with drain–source voltage.

Any further increase in V_{DS} cannot speed up electrons coming down the channel. The slight increase in I_{DS} with V_{DS} is due more to channel shortening.

There is an alternative view, that the conducting channel is never completely closed off by the depletion region, and the current passes at saturation velocity through a narrow thread of conducting channel. The net result is the same, effectively moving the drain nearer to the source with little change in I_{DS} above pinch off.

Inevitably, this region of operation is called saturation, as if the word were not already overburdened with different meanings. The *I/V* characteristic is shown in Figure 13.10, for different values of V_{GS}. The value of I_{DS} in saturation can be found by putting $V_{GS} - V_{DS} = V_p$ in equation (13.13), to eliminate V_{DS}:

$$I_{DS(sat)} = G_o\left[V_{GS} - V_p\left(\frac{1}{3} + \frac{2}{3}\left(\frac{V_{GS}}{V_p}\right)^{3/2}\right)\right]$$

(13.16)

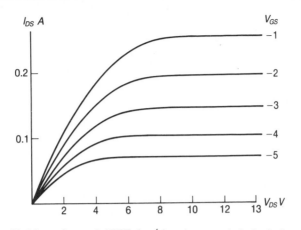

Figure 13.10 *n*-channel JFET I_{DS}/V_{DS} characteristic including the pinch off saturation region

Note that V_{GS} can only take the value $V_p + V_{DS}$ in this equation and $V_{DS} < V_p$.

In practice, non uniform doping and short channel effects make real JFETs follow a simpler empirical law in saturation:

$$I_{DS(sat)} \simeq I_{DSS}\left(1 - \frac{V_{GS}}{V_p}\right)^2 \tag{13.17}$$

where I_{DSS} is $I_{DS(sat)}$ with $V_{GS} = 0$.

EXAMPLE

■ Compare $I_{DS(sat)}$ from the two equations above for $V_{GS} = V_p$ and $V_{GS} = 0.5V_p$. Find the corresponding values of V_{DS} at saturation.

☐ First find I_{DSS} from equation (13.16) by putting $V_{GS} = 0$:

$$I_{DSS} = -G_o V_p/3$$

Now put $V_{GS} = V_p$ and $0.5V_p$ into equations (13.16) and (13.17) by turn:

1. $V_{GS} = V_p$

   (13.16):

   $$I_{DS(sat)} = G_o\left[V_p - \frac{V_p}{3}(1 + 2)\right]$$

   $$= 0$$

(13.17):

$$I_{DS(sat)} = \frac{-G_o V_p}{3} \left(1 - \frac{V_p}{V_p}\right)^2 = 0$$

If $V_{GS} = V_p$, then $V_{DS} = V_{GS} - V_p = 0$, so there is no voltage gradient along the channel and no drain–source current.

2. $V_{GS} = 0.5 V_p$

(13.16):

$$I_{DS(sat)} = G_o\left(\frac{V_p}{2} - \frac{V_p}{3}\left(1 + 2\left(\frac{1}{2}\right)^{3/2}\right)\right)$$

$$= -0.069 G_o V_p$$

(13.17):

$$I_{DS(sat)} = \frac{-G_o V_p}{3}\left(1 - \frac{0.5 V_p}{V_p}\right)^2$$

$$= -0.083 G_o V_p$$

In this case $V_{DS} = V_{GS} - V_p = -0.5 V_p$. V_p is negative, so V_{DS} and $I_{DS(sat)}$ are positive. The currents derived from the two models are similar but not identical.

13.4 Small-signal ac models

In normal operation we are concerned with the effect of fluctuations in gate–source and drain–source voltages (the inputs, especially V_{GS}) on the drain–source current (the output). As in the case of the bipolar junction transistor, small-signal quantities are designated by lower case symbols:

i_{ds}: small-signal ac component of I_{DS}
v_{ds}: small-signal ac component of V_{DS}
v_{gs}: small-signal ac component of V_{GS}

Since I_{DS} depends on both voltages:

$$I_{DS} = f(V_{GS}, V_{DS}) \tag{13.18}$$

a small change ΔI_{DS} in I_{DS} due to small changes ΔV_{GS} and ΔV_{DS} is given by:

$$\Delta I_{DS} = \left.\frac{\partial I_{DS}}{\partial V_{GS}}\right|_{V_{DS}=\text{constant}} \Delta V_{GS} + \left.\frac{\partial I_{DS}}{\partial V_{DS}}\right|_{V_{GS}=\text{constant}} \Delta V_{DS}$$

Replacing the small changes by their ac small-signal values (see section 12.2), this may be written:

$$i_{ds} = g_m v_{gs} + \frac{1}{r_{ds}} v_{ds}$$

where g_m is the *mutual conductance* of a JFET:

$$g_m = \frac{i_{ds}}{v_{gs}}\bigg|_{v_{ds}=0} = \frac{\partial I_{DS}}{\partial V_{GS}}\bigg|_{V_{DS}=\text{constant}} \qquad (13.19)$$

and r_{ds} is the small-signal *slope resistance* of the channel:

$$r_{ds} = \frac{v_{ds}}{i_{ds}}\bigg|_{v_{gs}=0} = 1\bigg/\frac{\partial I_{DS}}{\partial V_{DS}}\bigg|_{V_{GS}=\text{constant}} \qquad (13.20)$$

These expressions can be evaluated both before and after pinch off.

1. *Before pinch off*: the expression derived for I_{DS} as a function of V_{DS} and V_{GS} in equation (13.13) can be differentiated partially with respect to V_{GS} or V_{DS}. Here we will just provide the latter:

    $$\frac{\partial I_{DS}}{\partial V_{DS}} = G_o\left[1 - \left(\frac{V_{GS} - V_{DS}}{V_p}\right)^{1/2}\right] \qquad (13.21)$$

    If the gate and drain are connected externally, $V_{GS} = V_{DS}$ and:

    $$r_{ds} = 1/G_o \quad (V_{GS} = V_{DS}) \qquad (13.22)$$

    The same result can be reached by putting $V_{DS} \to 0$ in equation (13.13), which reduces to:

    $$I_{DS} = G_o V_{DS} \quad (V_{DS} \to 0) \qquad (13.23)$$

    Both these conditions mean that the gate–channel diode is not reverse biased, and the full channel is open for conduction. Beyond these restraints equation (13.21) implies that the small-signal channel resistance depends on the value of V_{GS}. This means that the JFET below pinch off behaves like a voltage-controlled resistor (VCR), a very useful circuit element. (We have cheated a little by obtaining $\partial I_{DS}/\partial V_{DS}$ at constant V_{GS}, and then allowing V_{GS} to vary. The resulting VCR still works.)

    The small-signal equivalent circuit is given in Figure 13.11a. The current i_{ds} depends on both V_{GS} and V_{DS}, because the precise values of both r_{ds} and g_m depend on the pair of voltages.

2. *After pinch off*: in saturation the current varies very little with drain–source voltage. This is equivalent to saying that $\partial I_{DS}/\partial V_{DS} \to 0$ or $r_{ds} \to \infty$. The small-signal equivalent circuit is therefore simplified (Figure 13.12b).

    This looks like the simplest ac model of a bipolar junction transistor,

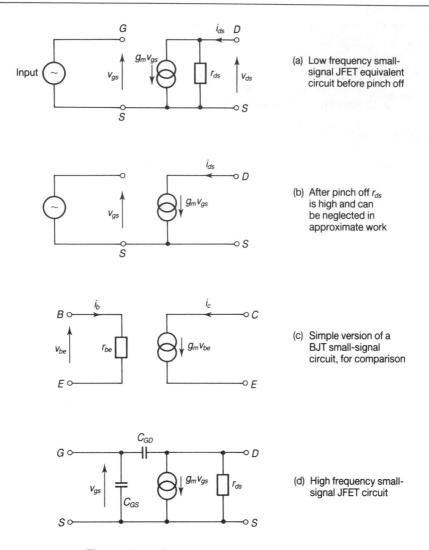

Input ~

(a) Low frequency small-signal JFET equivalent circuit before pinch off

(b) After pinch off r_{ds} is high and can be neglected in approximate work

(c) Simple version of a BJT small-signal circuit, for comparison

(d) High frequency small-signal JFET circuit

Figure 13.11 Small-signal equivalent circuit

repeated in Figure 13.12c, except that the JFET has a high input impedance and is voltage controlled. The BJT has a moderate input impedance and its output current is current controlled.

The value of g_m beyond pinch off can best be obtained from the empirical law, equation (13.17). In saturation:

$$\frac{\partial I_{DS}}{\partial V_{DS}} = \frac{-2I_{DSS}}{V_p}\left(1 - \frac{V_{GS}}{V_p}\right) = g_m \tag{13.24}$$

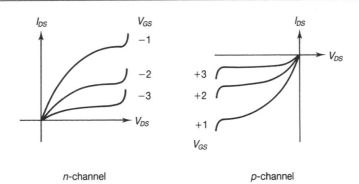

n-channel   *p*-channel

Note:
The curves are only well defined while the gate – channel diode is reverse biased. Breakdown occurs at high V_{DS}.

Figure 13.12 *n*-channel and *p*-channel JFET characteristics

The JFET is not therefore a wholly linear device, because the transconductance depends on the value of input voltage.

At high frequencies, junction capacitances must be added to the model, as in Figure 13.12d. The gate–source and gate–drain capacitances are dominant, with typical values of a few picofarads for discrete devices.

13.5 Real JFETs

All the analysis above has been for *n*-channel JFETs, but *p*-channel JFETs also exist. The I/V characteristics of the two types are contrasted in Figure 13.12. *p*-channel JFETs have a higher channel resistance and a lower top frequency because hole mobility and saturation velocity are both lower than those of electrons.

I_{DS} is not truly constant in saturation, but increases slightly with V_{DS} because of shortening of the channel. An output admittance g_o can be defined:

$$g_o = \left| \frac{\partial I_{DS}}{\partial V_{DS}} \right|_{V_{GS}=\text{constant}} \quad (= 1/r_{ds}) \quad (13.25)$$

It has a typical value of 25 μA V^{-1}.

At high values of V_{DS} both types exhibit breakdown, in the form of a sharp rise in I_{DS}. This is caused by avalanche multiplication of mobile charges in the high field region of the channel. Breakdown need not be destructive if the external circuit limits I_{DS}.

The internal structure of real JFETs differs from that given previously

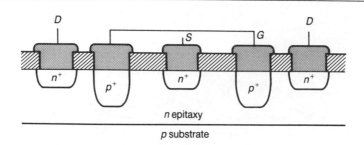

Figure 13.13 Real JFET cross-section

because of the difficulty in creating the gate junction vertically below the channel. A typical cross-section is given in Figure 13.13.

13.6 MESFETs

Metal–semiconductor FETs have the same general characteristics as JFETs, except that a Schottky barrier diode replaces the junction diode between the gate and the channel. MESFETs find particular application in III–V semiconductors such as GaAs where the high electron mobility leads to short switching times. Further, the possibility of making a MESFET without diffusing dopants allows much smaller devices to be made, again with a reduction in switching time (Figure 13.14). In this case the high frequency gain may not be limited by the transit time of majority carriers between source and drain, but by the time constant associated with charging the gate capacitance.

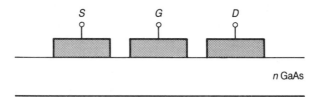

Figure 13.14 GaAs MESFET structure

EXAMPLE

■ Find the mutual conductance of a JFET in saturation with a pinch off voltage of -8 V, a gate–source voltage of -4 V, if $I_{DSS} = 20$ mA. Find the small signal voltage gain if the load resistance is 1,000 Ω.

☐   The mutual conductance in saturation is given by equation (13.24):

$$g_m = \frac{-2 \times 0.02}{-8}\left(1 - \frac{4}{8}\right) = 2.5 \times 10^{-3} \text{ A V}^{-1}$$

If an input ac signal v_{gs} is applied to the gate, the ac drain–source current will be:

$$i_{ds} = g_m v_{gs}$$

This current flows through a 1,000 Ω load resistor R_L and creates an ac output voltage v_o:

$$v_o = i_{ds}R_L = g_m v_{gs} R_L$$

The voltage gain is therefore A_V:

$$A_V = \frac{v_o}{v_{gs}} = g_m R_L = 2.5$$

This analysis neglects the variation of g_m with V_{GS} and is restricted to small signals ($|v_{gs}| \ll |V_{GS}|$).

Summary

Main symbols introduced

Symbol	Unit	Description
a	m	half the channel height
g_o	A V^{-1}	output admittance
g_m	A V^{-1}	mutual conductance
G_o	S	channel conductance
I_{DSS}	A	drain–source current at saturation with $V_{GS} = 0$
L	m	channel length
r_{ds}	Ω	small-signal resistance of channel; on resistance
V_p	V	gate–channel voltage which pinches off the channel
w	m	gate–channel depletion region width on each side
z	m	channel width perpendicular to the drain–source current

Subscripts

CH	channel	G, g	gate
D, d	drain	S, s	source

Main equations

All equations were given for an n-channel JFET, for which:

$$V_{DS} > 0, \; V_{GS} < 0 \; V_p < 0$$

$$I_{DS} = G_o\left[V_{DS} + \frac{2}{3}V_p\left\{\left(\frac{V_{GS} - V_{DS}}{V_p}\right)^{3/2} - \left(\frac{V_{GS}}{V_p}\right)^{3/2}\right\}\right] \quad \text{(before pinch off)} \quad (13.13)$$

$$G_o = 2zaqN_D\mu_e/L \tag{13.11}$$

$$V_p = -qa^2N_D/2\epsilon_0\epsilon_r \tag{13.12}$$

$$I_{DS(sat)} \approx I_{DSS}\left(1 - \frac{V_{GS}}{V_p}\right)^2 \quad \text{(after pinch off)} \tag{13.17}$$

$$g_m = \left.\frac{\partial I_{DS}}{\partial V_{GS}}\right|_{V_{DS}=\text{constant}} = \left.\frac{i_{ds}}{v_{gs}}\right|_{v_{ds}=0} \tag{13.19}$$

$$r_{ds} = 1\left/\left.\frac{\partial I_{DS}}{V_{DS}}\right|_{V_{GS}=\text{constant}}\right. = \left.\frac{v_{ds}}{i_{ds}}\right|_{v_{gs}=0} \tag{13.20}$$

$$\approx 1/G_o \text{ at low } I_{DS}$$

$$\approx \infty \text{ in saturation}$$

Exercises

1. Sketch cross-sectional diagrams for both n-channel and p-channel JFETs, indicating the polarities of V_{GS} and V_{DS} in normal operation. Is it possible to construct a JFET for which $I_{DS} = 0$ for $V_{GS} = 0$?

2. For a JFET to have the highest sensitivity to voltage signals applied to the gate, should the channel or the gate region be more heavily doped? Explain your answer.

3. An n-channel silicon JFET is 100 microns wide, has a 5 micron channel length and a half-channel height of 0.5 microns. What doping density will give a minimum channel resistance of 0.8 Ω?

4. Find the pinch off voltage for the JFET in (3). If $|V_p|$ is reduced how will this affect the channel resistance?

5. A p-channel silicon JFET has a full channel height of 3 microns, channel length 8 microns and width 400 microns. The channel doping density is 10^{21} m^{-3} and the gate doping is much higher. Find the pinch off voltage and the drain–source saturation current for $V_{GS} = 1$ V. What is V_{DS} at the onset of saturation in this case?

6. Sketch band diagrams for a *p*-channel JFET:
   (a) a vertical cross-section through the gate and channel before pinch-off;
   (b) a horizontal section along the channel after pinch off.

7. An *n*-channel silicon JFET has a channel length of 2 microns, channel width 0.3 mm, pinch off voltage of -10 V and an on resistance of 120 Ω. Find the channel doping density and the full channel height.

8. Show that g_m for an *n*-channel JFET before pinch off is given by:
   $$g_m = G_o\left[\left(\frac{V_{GS} - V_{DS}}{V_p}\right)^{1/2} - \left(\frac{V_{GS}}{V_p}\right)^{1/2}\right]$$
   Is g_m positive or negative?
   Find the maximum and minimum values of g_m within the range of validity.

9. An *n*-channel JFET has a pinch off voltage of -18 V and $I_{DSS} = 0.2$ A. Find g_m in saturation at $V_{GS} = -5$ V. What is the peak to peak ac signal which can be superimposed on V_{GS} if g_m is to remain within 10% of its nominal dc value?

10. Superimpose on Figure 13.10 a load line for a supply voltage of 12 V and a load resistance of 60 Ω. Hence find the swing in V_{DS} in response to a switch in V_{GS} from -1 V to -5 V.

14 The MOS capacitor

14.1 Description

The three major classes of field effect transistor are the junction FET (JFET), the metal–semiconductor FET (MESFET) and the metal–insulator–semiconductor FET (MISFET). By far the most important insulating MISFET film is silicon dioxide (oxide) leading to the name MOSFET.

The purpose of this chapter is to explore the properties of the core of the MOSFET, which is a sandwich of metal, silicon dioxide insulator, and silicon semiconductor, as shown in Figure 14.1. The sandwich behaves electrically like a capacitor, because dc current can only pass through the metal and the semiconductor, and is called an MOS capacitor. It can be used as an integrated capacitor but that is not its primary use. The semiconductor part of it will form the channel of the MOSFET, while the metal will be the gate connection. Note that in most of the diagrams of the MOS capacitor it will be turned on its side, compared to the cross-section of the MOSFET. This is because the band diagram of a *vertical* cross-section through the gate insulator and semiconductor will be of most interest in examining the electrical characteristics.

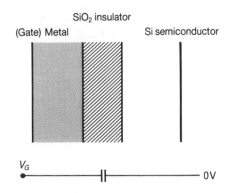

Figure 14.1 Metal-oxide-semiconductor (MOS) capacitor

In subsequent sections the band diagram will be derived for various conditions of bias, namely:

1. *Zero bias*.
2. *Accumulation*, where the gate potential attracts majority carriers in the channel.
3. *Depletion*, where the gate potential repels majority carriers in the channel.
4. *Inversion*, where attracted minority carriers exceed equilibrium majority carriers at the interface.

These cases will be considered in more detail below.

The fabrication of an MOS capacitor is simple to describe, but difficult to accomplish well. The steps are as follows:

1. Dope the semiconductor to the required type and sheet resistance.
2. Grow a silicon dioxide layer, typically 10–100 nm thick, usually by a dry oxidation process.
3. Deposit a metal layer, such as aluminium.

In practice the doping is trimmed after growing the oxide layer by implanting dopants *through* the thin oxide. A good MOS capacitor requires the most scrupulous cleanliness in all stages of preparation. A mistimed sneeze or (perish the thought) a fingerprint would ruin not only the wafers concerned but subsequent batches oxidized in the same furnace as the dirty wafers.

14.2 Zero bias

We consider the case of a p-type semiconductor connected to ground. The gate–semiconductor potential difference is then simply the gate potential V_G

with respect to ground. In this ideal analysis we will neglect charged interface states and charged impurity ions other than dopants. It will also be assumed that the metal and semiconductor work functions ϕ_m and ϕ_s are identical. Both these unwarranted assumptions will be relaxed later.

The band diagram for an MOS capacitor with zero bias is given in Figure 14.2. The Fermi level is horizontal, indicating equilibrium. The band edges are level, meaning zero current flow. The bump in the vacuum level is due to the contact potentials of metal and semiconductor with SiO_2. The vacuum level is equal on either side of the SiO_2 because we have assumed $\phi_m = \phi_s$. The very large ($\simeq 8$ eV) band gap in SiO_2 ensures that mobile charges are presented with large potential barriers and are most unlikely to cross, i.e. it acts as an insulator.

Note the appearance of other quantities:

1. ϕ'_m and ϕ'_s (eV) are the *modified work functions* for the metal and semiconductor respectively. These are measured from the Fermi level E_F to the SiO_2 conduction band edge. In later diagrams the vacuum level will be omitted and the modified work functions will be used. There is no particular reason for this, other than simplicity of diagrams and to keep an eye on the potential barriers faced by mobile charges. (The Fermi level is taken as the reference level here because it is flat. This is equivalent to making $\phi = 0$ in equations (4.14)ff.)

2. V_F is the separation in electron-volts between E_F and the intrinsic Fermi level E_{Fi}. V_F (V_F in this book is equivalent to ϕ_F in several others.) is positive in a *p*-type semiconductor, defined by:

$$qV_F = E_{Fi} - E_F \text{ J} \qquad (14.1)$$

The factor q (>0) converts from electron-volts to energies in joules. V_F is

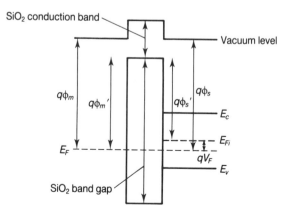

Figure 14.2 MOS capacitor ideal band diagram $V_G = 0$

a measure of how strongly the semiconductor is doped. Recalling equation (4.12):

$$p = n_i \exp \frac{E_{Fi} - E_F}{kT} = n_i \exp \frac{qV_F}{kT} \tag{14.2}$$

With $V_F > 0$, the exponent is greater than one and holes are in the majority. This quantity will be used in defining the inversion condition.

14.3 Accumulation ($V_G < 0$, p-type)

If a negative potential is applied to the gate, positively charged holes in the semiconductor are attracted towards the metal and accumulate there.

The quantity of charge per unit area depends on the capacitance of the oxide, C_{ox}, and the voltage applied. The capacitance is given by:

$$C_{ox} = \frac{\epsilon_{ox}}{t_{ox}} \, \mathrm{F\,m^{-2}} \tag{14.3}$$

where $\epsilon_{ox} \, \mathrm{F\,m^{-1}}$ is the permittivity of SiO_2 and t_{ox} its thickness. The magnitude of the charge on each side of the insulator Q is:

$$Q = |V_G C_{ox}| \, \mathrm{C\,m^{-2}} \tag{14.4}$$

In accumulation a p-type semiconductor has an excess of holes at the interface, while an equal excess of electrons exists at the metal–semiconductor interface. Both charge sheets are very thin because of the large number of available states in the two materials. The capacitor is well represented by two thin charge sheets separated by the SiO_2 thickness.

Nearly all the voltage is dropped across the insulating film. This is because the volt drop V across a charge sheet depends on both the charge per unit volume, c, and the thickness, d, of the sheet. The solution of Poisson's equation for such a sheet was given in Chapter 1. The volt drop is:

$$V = \frac{cd^2}{2\epsilon} \tag{14.5}$$

The product cd is simply Q, the charge per unit area on the capacitor. The charge density c is of order qN_v, where N_v is the density of available valence band states – about $10^{25} \, \mathrm{m^{-3}}$ in silicon. Using equations (14.3) and (14.4), we obtain for the fraction of the gate voltage dropped across the accumulation layer:

$$\frac{V}{V_G} = \frac{V_G \epsilon_{ox}{}^2}{2\epsilon q N_v t_{ox}{}^2} \tag{14.6}$$

For a 50 nm gate oxide and 5 V gate potential, $V/V_G = 0.007$. That is, less than

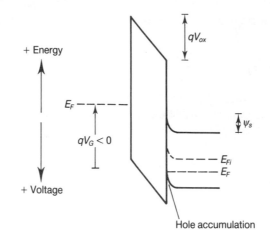

Note:.
The gate metal is negative with respect to the semiconductor. The semiconductor side of the capacitor has a positive charge of excess holes near the surface.

Figure 14.3 p-type MOS capacitor in accumulation

1% of the total voltage is dropped across the semiconductor. Similar considerations apply at the metal–insulator interface.

Taking the MOS capacitor into accumulation has not altered it drastically. The insulator–semiconductor interface region has become more conducting, but it remains p-type. The band diagram in Figure 14.3 summarizes the condition:

1. The Fermi level is raised on the metal side, for a negative gate potential.
2. The tight downward curvature at the oxide–semiconductor interface indicates an accumulation of positive charge.
3. Horizontal Fermi levels and band edges away from the interfaces mean no current flows.
4. Steeply sloping oxide band edges indicate a strong electric field.

14.4 Depletion ($V_G > 0$, p-type)

A positive gate potential applied to the metal tends to drive away positive holes in the semiconductor, leaving the negatively charged acceptor dopants uncompensated. This creates a depletion region at the insulator–semiconductor interface. The charge density in the depletion region is $qN_A\,\mathrm{C\,m^{-3}}$, assuming all holes are repelled, which is generally much lower than qN_v for the accumulation region. The same expressions for voltage drop across the charge sheet now apply, with N_A replacing N_v.

We cannot now assume that all the voltage is dropped across the insulator. Instead a surface potential ψ_s is set up at the semiconductor interface, such that:

$$V_G = \psi_s + V_{ox} \tag{14.7}$$

where V_{ox} is the voltage across the oxide. The band diagram for this case is given in Figure 14.4. Holes float away from the interface, leaving a negatively charged layer denoted by the upwards curvature. The separation in Fermi levels across the device is the sum of the potential drop across the SiO$_2$ and ψ_s across the depletion region.

The charge in the depletion region Q_d Cm^{-2} must then equal the charge induced by V_{ox} on each plate of the capacitor. The relation between depletion region width d and surface potential ψ_s is (rearranging (14.5)):

$$d = \left(\frac{2\epsilon_s \psi_s}{q N_A}\right)^{1/2} \tag{14.8}$$

ϵ_s is the semiconductor permittivity in Fm^{-1} Q_d is given by:

$$|Q_d| = |-q N_A d|\ \mathrm{C\,m^{-2}} \tag{14.9}$$

and the charge Q_d supported by V_{ox} is:

$$|Q_d| = V_{ox} C_{ox} \tag{14.10}$$

Equations (14.7)–(14.10) then form a set of equations with d, Q_d, V_{ox} and ψ_s as unknowns. (If the solution appears tedious to the student, reflect on the marvellous way every MOS capacitor solves it correctly in milliseconds!)

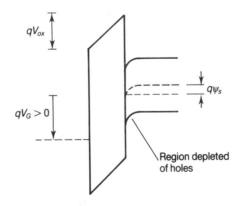

Note:
The semiconductor side of the capacitor has a negative depletion region charge.

Figure 14.4 *p*-type MOS capacitor in depletion: $V_G > 0$

EXAMPLE

■ Find the surface potential and depletion region width for a p-type silicon MOS capacitor doped 10^{23} m^{-3} with an oxide thickness of 0.1 μm if the gate potential is 5 V.

☐ Eliminating V_{ox}, Q_d and d between equations (14.7)–(14.10) a quadratic equation in ψ_s is formed:

$$\psi_s^2 - \psi_s\left(2V_G + \frac{2\epsilon_s qN_A}{C_{ox}^2}\right) + V_G^2 = 0 \tag{14.11}$$

Inserting the data given, with $C_{ox} = \epsilon_{ox}/t_{ox}$ F m^{-2}, and relative permittivities of 3.9 and 11.9 for SiO$_2$ and silicon respectively:

$$\psi_s^2 - 38.29\psi_s + 25 = 0$$

with a solution:

$$\psi_s = 0.22 \text{ V}$$

The other root has $\psi_s > V_G$, which is non-physical.
    Thus $V_{ox} = V_G - \psi_s = 4.34$ V and:

$$|Q_d| = C_{ox}V_{ox} = \frac{V_{ox}C_{ox}}{t_{ox}} = 1.5 \times 10^{-3} \text{ C m}^{-2}$$

Thus the depletion region width is:

$$d = Q_d/qN_A = 9.35 \times 10^{-8} \text{ m} \tag{14.12}$$

In this case the charge is spread out over a similar thickness to the insulator thickness, so the solution is only approximate. The use of the oxide capacitance assumes thin charge sheets spaced by the SiO$_2$ dielectric film. In a more accurate approach, C_{ox} would be reduced, Q_d and ψ_s would fall and V_{ox} would rise.

14.5 Inversion ($V_G \gg 0$, p-type)

Suppose that the gate voltage is taken increasingly positive. First, the depletion region expands as the charge on the MOS capacitor grows. At the same time the intrinsic Fermi level E_{Fi} is depressed more and more, with the greatest lowering at the interface. While E_{Fi} remains above the Fermi level E_F, equation (14.2) predicts that the hole density will be greater than the intrinsic carrier density and the material can be considered p-type in spite of the depletion.
    As V_G is increased there comes a point at which $E_{Fi} = E_F$. The minority

carrier electrons which tend to gather in the depressed region have been neglected so far but from now on they must be considered. A further increase in V_G will take E_{Fi} below E_F (Figure 14.5) and electrons become the majority carriers (equation (4.11)):

$$n = n_i \exp \frac{E_F - E_{Fi}}{kT} \tag{14.13}$$

When $E_F > E_{Fi}$, $n > n_i$ and the surface layer is *inverted* to become *n*-type! This is a remarkable phenomenon, quite unlike the channel *dimension* control seen in the JFET. An *n*-type channel can be called into existence in *p*-type material, simply by applying a voltage the other side of an insulator.

The device capabilities of an MOS capacitor will be explored in the next chapter. First we must find how much voltage must be applied to the gate to invert the semiconductor. This is known as the *threshold voltage, V_{TH}*.

One common, although arbitrary, definition of a useful amount of inversion is:

The interface and substrate carrier densities must be equal.

That is, the inversion layer must be as *n*-type as the substrate is *p*-type. This implies that E_{Fi} must be depressed below E_F by an amount qV_F, as given in

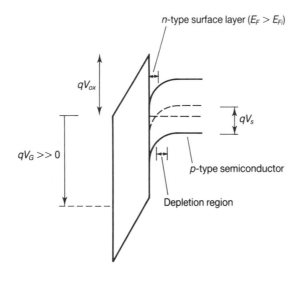

Note:
The negative charge at the semiconductor surface consists of a depletion region and a thin *n*-type 'inversion' layer.

Figure 14.5 *p*-type MOS capacitor biased to inversion: $V_G \gg 0$

equations (14.1) and (14.2). The surface potential ψ_s at threshold condition is therefore:

$$\psi_s = 2V_F \quad \text{(threshold)} \tag{14.14}$$

The band diagram of the MOS capacitor at threshold in Figure 14.5 will repay study to make this point clear. Bear in mind that the threshold condition is the *onset* of useful inversion: higher gate voltages will render the channel increasingly *n*-type, by bending the bands even further.

If it can be assumed that the inversion electron charge per unit area Q_n is much smaller than the depletion charge Q_d, the same formulae can be applied as those in the previous section. The total gate voltage at threshold V_{TH} is the sum of the voltage drops across the oxide and the semiconductor:

$$V_{TH} = V_{ox} + \psi_s \quad \text{(threshold)}$$

$$= \frac{Q_d}{C_{ox}} + 2V_F \tag{14.15}$$

The charge Q_d is known because the potential supporting it as a depletion region is $2V_F$. Using equations (14.8) and (14.9):

$$|Q_d| = qN_A d$$

$$= (4\epsilon_s q N_A V_F)^{1/2} \, \text{C}\,\text{m}^{-2} \quad \text{(threshold)} \tag{14.16}$$

V_F can be found from the value of p in the unperturbed semiconductor:

$$p = N_A = n_i \exp \frac{qV_F}{kT} \tag{14.17}$$

Hence:

$$V_F = \frac{kT}{q} \ln \frac{N_A}{n_i} \tag{14.18}$$

All terms are now known in equation (14.15), and the threshold voltage can be calculated.

EXAMPLE

■ Find the threshold voltage for an ideal silicon MOS capacitor if $t_{ox} = 80$ nm, $N_A = 10^{23}$ m^{-3} and the temperature is 300 K.

□ Using equation (14.18):

$$V_F = \frac{1.38 \times 10^{-23} \times 300}{1.6 \times 10^{-19}} \ln \frac{10^{23}}{1.5 \times 10^{16}}$$

$$= 0.41 \text{ V}$$

The depletion region charge with a surface potential $2V_F$ follows from equation (14.16):

$$|Q_d| = (4 \times 11.9 \times 8.85 \times 10^{-12} \times 1.6 \times 10^{-19} \times 10^{23} \times 0.41)^{1/2}$$

$$= 1.66 \times 10^{-3}\,\mathrm{C\,m^{-2}}$$

The oxide capacitance per unit area is given by equation (14.3):

$$C_{ox} = \frac{3.9 \times 8.85 \times 10^{-12}}{8 \times 10^{-8}}$$

$$= 4.31 \times 10^{-4}\,\mathrm{F\,m^{-2}}$$

These quantities can be inserted in equation (14.15):

$$V_{TH} = \frac{1.66 \times 10^{-3}}{4.31 \times 10^{-4}} + 2 \times 0.41$$

$$= 3.85 + 0.82$$

$$= 4.67\,\mathrm{V}$$

An n-type layer exists in the p-type semiconductor for gate voltages of 4.67 V and higher, with respect to the bulk of the semiconductor.

14.6 Strong inversion ($V_G > V_{TH}$, p-type)

When the gate voltage is taken beyond threshold, the interface region becomes more and more strongly n-type.

Increases in the negative charge stored are now due to more conduction band electrons, rather than an increase in the depletion region thickness. A good approximation is to say that the threshold voltage supports the depletion region charge Q_d $\mathrm{C\,m^{-2}}$ and $(V_G - V_{TH})$ holds the mobile electron charge, Q_n $\mathrm{C\,m^{-2}}$:

$$Q_d = V_{TH}C_{ox} \quad \text{(strong inversion)}$$

$$Q_n = (V_G - V_{TH})C_{ox} \tag{14.19}$$

The MOSFET described in the next chapter relies on controlling Q_n by the gate voltage, in order to influence the current flowing along the inversion layer.

It is instructive to look at the charge distribution in the capacitor for the different bias conditions. Figure 14.6 illustrates this for a p-type MOS capacitor in accumulation, depletion, and strong inversion. They should be studied together with the relevant band diagrams. The student should be able to reproduce these drawings, preferably relying on understanding rather than memory.

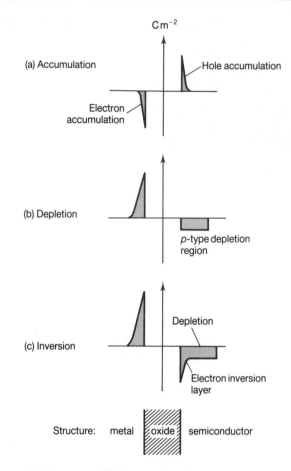

Figure 14.6 p-type MOS capacitor charge distribution

14.7 Real MOS capacitors

The ideal MOS capacitor treated above neglects several physical phenomena which are nearly always present to a significant degree. This section lists the most important sources of error, and indicates how to include them in the expression for threshold voltage.

1. *Interface trapped charge Q_i Cm^{-2}*: the regular crystal lattice in the bulk of the silicon is interrupted at the interface with the silicon dioxide film, because SiO_2 is an amorphous material and lacks long-range order. A

bare, clean silicon surface has a Q_i of order $1 \, C \, m^{-2}$, equivalent to one electronic charge per atom in the surface layer. A good SiO_2 layer, annealed in hydrogen at $450 \, °C$ can have a Q_i value as low as $10^{-5} \, C \, m^{-2}$. A low value is important because interface trapped charge resides in levels within the silicon band gap, which are also fast recombination centres. A high surface recombination velocity rapidly soaks up mobile carriers and prevents the creation of an inversion layer.

MOSFETs are nearly always constructed on $\langle 100 \rangle$ silicon surfaces, because these contain the smallest number of available silicon bonds and have the lowest Q_i values.

2. *Fixed oxide charge $Q_f \, C \, m^{-2}$*: a layer of charge also exists within the oxide film close to the SiO_2/Si interface. Its value varies little with oxide thickness or semiconductor doping and may arise from a strained layer of SiO_2. The charge is nearly always positive.

3. *Trapped oxide charge $Q_t \, C \, m^{-2}$*: defects within the SiO_2 layer are usually neutral, but can become charged, for example by a high voltage pulse applied to the capacitor. This can be useful, and is the basis of one memory device (see Chapter 15). In the simple MOSFET it needs to be minimized if threshold voltages are to be kept reproducible and low.

4. *Mobile ionic charge $Q_m \, C \, m^{-2}$*: certain impurities, notably the alkali metals sodium and potassium, can be present in the form of positively charged ions. They can move about within the silicon dioxide film in response to applied voltages, especially at elevated temperatures, leading to a gradual drift in the threshold voltage. This instability limited the early progress in MOSFET technology: only when scrupulous attention was paid to cleanliness were reliable devices fabricated routinely.

5. *Work function difference $\phi_{ms} \, eV$*: it was assumed for simplicity that the metal work function ϕ_m was equal to the semiconductor work function ϕ_s. However, to take a typical example, $\phi_m = 4.1 \, eV$ in aluminium and $\phi_s = 4.6 \, eV$ in intrinsic silicon. Further, the semiconductor work function varies with the doping density. We define:

$$\phi_{ms} = \phi_m - \phi_s \, eV \tag{14.20}$$

as the difference between the two.

Now we consider how the threshold voltage is altered by these effects. For simplicity it is assumed that all the charge components (1)–(4) reside close to the oxide/semiconductor interface, where they can be lumped together as Q_o:

$$Q_o = Q_i + Q_f + Q_t + Q_m \, C \, m^{-2} \tag{14.21}$$

This charge is nearly always positive. It induces an opposite charge within the semiconductor, as shown in Figure 14.7a. In fact some charge is also induced at the metal–insulator interface due to the finite spreads of Q_o: this will be neglected here. The negative charge induced in the semiconductor appears as a depletion region in p-type material.

Charges are also induced by the work function differences. For example, if ϕ_{ms} is negative, the metal Fermi level is lowered and the semiconductor E_F is raised when the capacitor is made, until the Fermi levels coincide in equilibrium.

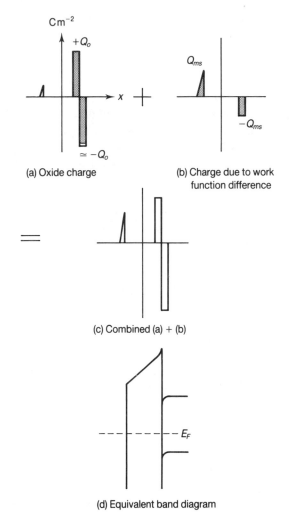

(a) Oxide charge

(b) Charge due to work function difference

(c) Combined (a) + (b)

(d) Equivalent band diagram

Figure 14.7 Charge distribution in a MOS capacitor

This requires a positive charge on the metal side and a negative charge in the semiconductor, both as thin charge sheets of value $\pm Q_{ms}$ $\mathrm{C\,m}^{-2}$, where:

$$Q_{ms} = \phi_{ms} C_{ox} \ \mathrm{C\,m}^{-2} \tag{14.22}$$

Charges due to ϕ_{ms} are illustrated in Figure 14.7b. The two sets of charges add as in Figure 14.7c, giving rise to the band diagram in Figure 14.7d in equilibrium.

Comparing this with the ideal band diagrams for equilibrium and depletion (Figures 14.2 and 14.4), it is clear that the p-type semiconductor is part way to inversion, without any voltage being applied. In order to restore the oxide and semiconductor to the ideal field-free condition it would be necessary to apply a voltage V_{fb} to the metal side, with respect to the grounded semiconductor, where:

$$V_{fb} = -(Q_o + Q_{ms})/C_{ox}$$

$$= \phi_{ms} - \frac{Q_o}{C_{ox}} \tag{14.23}$$

for a negative ϕ_{ms} and a positive Q_o. V_{fb} is called the flat-band voltage because it removes the curvature from the semiconductor band diagram, as in Figure 14.8.

The threshold voltage $V_{TH'}$ for the onset of strong inversion in the non-ideal case is altered by the same amount:

$$V_{TH'} = \frac{Q_d}{C_{ox}} + 2V_F + V_{fb} \tag{14.24}$$

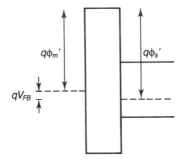

Note:
This brings the semiconductor band diagram to the ideal condition.

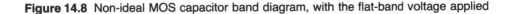

Figure 14.8 Non-ideal MOS capacitor band diagram, with the flat-band voltage applied

The first two terms are positive in a p-type substrate and negative in an n-type substrate. If in doubt work at the band diagram for inversion. The voltage drops across the oxide and across the semiconductor depletion region reinforce each other.

The flat-band voltage is usually negative. The threshold voltage for a non-ideal p-type MOS capacitor is therefore reduced below its ideal (positive) value. In the extreme case V_{fb} can exceed $(Q_d/C_{ox} + 2V_F)$, which makes the threshold voltage negative even in a p-type capacitor, and inversion layer will exist at $V_G = 0$.

For an n-type substrate V_{fb} reinforces the ideal (negative) terms, giving a more negative threshold voltage. Historically, this meant that MOS devices in n-type substrates were made first because inversion layers were unlikely to be formed accidently.

EXAMPLE

■ Find the threshold voltage for a silicon MOS capacitor with the following characteristics: $\phi_{ms} = -0.4$ V, $N_A = 10^{20}$ m^{-3}, $Q_o = 8 \times 10^{-5}$ Cm^{-2}, $t_{ox} = 100$ nm; oxide and silicon relative permittivities 3.9 and 11.9 respectively; $T = 300$ K.

□ First find the ideal threshold voltage. The oxide capacitance is:

$$C_{ox} = \frac{\epsilon_0 \epsilon_r}{t_{ox}} = \frac{8.85 \times 10^{-12} \times 3.9}{10^{-7}} = 3.45 \times 10^{-4} \, \mathrm{F \, m^{-2}}$$

The value of V_F is found using equation (14.18):

$$V_F = \frac{kT}{q} \ln \frac{N_A}{n_i} = \frac{1.38 \times 10^{-23} \times 300}{1.6 \times 10^{-19}} \ln \frac{10^{20}}{1.5 \times 10^{16}}$$
$$= 0.228 \text{ V}$$

Q_d is found from equation (14.16):

$$Q_d = (4 \times 8.85 \times 10^{-12} \times 11.9 \times 1.6 \times 10^{-19} \times 10^{20} \times 0.228)^{1/2}$$
$$= 3.92 \times 10^{-5} \, \mathrm{C \, m^{-2}}$$

The ideal threshold voltage is therefore:

$$V_{TH} = \frac{|Q_d|}{C_{ox}} + 2V_F$$
$$= 0.114 + 0.456$$
$$= 0.57 \text{ V}$$

The flat-band voltage is given by equation (14.23):

$$V_{fb} = \phi_{ms} - Q_o/C_{ox}$$

$$= -0.4 - \frac{8 \times 10^{-5}}{3.45 \times 10^{-5}}$$

$$= -2.72 \text{ V}$$

The non-ideal threshold voltage follows from equation (14.24):

$$V_{TH'} = V_{TH} + V_{fb}$$

$$= 0.57 - 2.72$$

$$= -2.15 \text{ V}$$

An *n*-type inversion layer is formed for gate–substrate voltages more positive than -2.15 V, for example at zero volts. The low doping density makes the substrate easy to invert, while the moderately high interface charge causes inversion unless a negative gate potential is applied to prevent it.

14.8 *C/V* measurement

The most widespread means of characterising an MOS capacitor is by means of a *C/V* plot. This is a graph of the *small-signal* capacitance against the dc voltage across the capacitor. It is only necessary to form two contacts, by means of probes, and to inject a small ac signal superimposed on a dc voltage, as part of a bridge measurement circuit. Automated test equipment exists for this purpose. Different curves are obtained for *p*- and *n*-type semiconductors, and for high (~ 1 MHz) and low (< 100 Hz) excitation frequencies.

14.8.1 *p-type (Figure 14.9a)*

With strongly negative gate bias the capacitor is in accumulation, so the charge sheets are thin and close to the insulator. This makes the small-signal capacitance close to the value for C_{ox} (multiplied by the area of the actual capacitor). Majority carriers are the only charges moving and respond well at all frequencies up to a few MHz. The high- and low-frequency curves lie together in this region.

As the dc bias is made more positive the capacitor goes towards inversion, with a depletion region being formed. As long as minority carriers are negligible

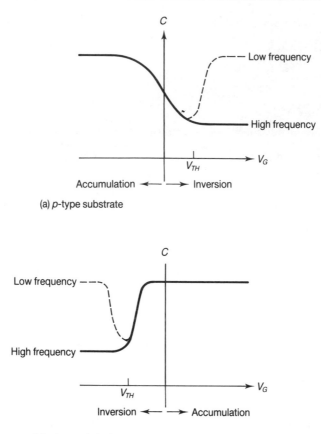

Figure 14.9 *C/V* curves for MOS capacitors

at the interface (i.e. in depletion), the change in charge storage is due to the depletion region expanding and contracting in response to the small ac signal. The MOS capacitor now consists of C_{ox} in series with the depletion region capacitance C_d (Figure 14.10):

$$C = \frac{C_d C_{ox}}{C_d + C_{ox}} \; \mathrm{F\,m}^{-2} \tag{14.25}$$

C_{ox} is fixed, but C_d varies according to the potential across the depletion region (the surface potential). C_d falls with an increase in gate voltage because of the rise in depletion region width, until the inversion region forms.

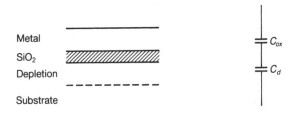

Figure 14.10 Oxide and depletion region capacitances

In strong inversion the depletion region width is nearly constant for different dc potentials. Different capacitances are obtained at high and low frequencies.

1. Low frequency ($< 100\,\text{Hz}$): electrons in the inversion layer can follow the ac signal and the small-signal capacitance reverts to C_{ox}, as in accumulation.

2. High frequency ($\gg 100\,\text{Hz}$): inversion layer carriers cannot change their number fast enough, and the depletion region width oscillates. The MOS capacitance is frozen at its minimum value.

14.8.2 n-type (Figure 14.9b)

The same behaviour is seen in an *n*-type MOS capacitor, except for a change of polarity. Positive gate voltages attract majority electrons to the interface, making the small-signal capacitance C_{ox}. Progressively decreasing V_G first creates a depletion region, reducing the capacitance to $C_{ox}C_d/(C_{ox} + C_d)$, then forms a hole-rich inversion layer. The capacitance is either constant or increases back to C_{ox}, depending on the excitation frequency.

The *C/V* characteristic can be used to check the doping density or oxide thickness, provided the other is known. Far more importantly the threshold voltage is found directly. In both *n*- and *p*-type, V_{TH} is where the *C/V* plot levels out in inversion. Automatic test equipment employs algorithms to find V_{TH} reproducibly from noisy traces. Alternatively, asymptotes can be judged by eye on the *C/V* plot, as in Figure 14.9.

A positive charge within the oxide causes a negative shift in V_{TH}, for both *n*- and *p*-type substrates. The entire *C/V* curve is displaced along the voltage axis, as shown in Figure 14.11.

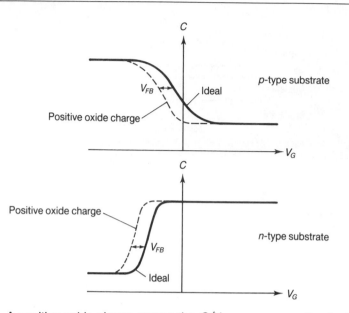

Figure 14.11 A positive oxide charge moves the C/V response negative by V_{FB} for both types of substrate

EXAMPLE

■ Find the ratio of minimum and maximum capacitances for the MOS capacitor in the last example.

□ The depletion region capacitance is the same as for a single-sided pn junction (from equation (7.9)):

$$C_d = \left(\frac{\epsilon q N_D}{4 V_F}\right)^{1/2} \mathrm{F\,m^{-2}} \tag{14.26}$$

where we have assumed that the maximum surface potential is $2V_F$, for the smallest capacitance. Hence:

$$C_d = \left(\frac{8.85 \times 10^{-12} \times 11.9 \times 1.6 \times 10^{-19} \times 10^{20}}{4 \times 0.228}\right)^{1/2}$$

$$= 4.30 \times 10^{-5} \mathrm{F\,m^{-2}}$$

while the oxide capacitance was previously found to be $3.45 \times 10^{-4} \mathrm{F\,m^{-2}}$. Equation (14.25) gives the value of these components in series:

$$C = \frac{C_d C_{ox}}{C_d + C_{ox}} \mathrm{F\,m^{-2}}$$

This has its minimum value when C_d is smallest, in strong inversion. It is always smaller than C_{ox}, the accumulation capacitance. The ratio of minimum and maximum capacitance is therefore:

$$\frac{C_{min}}{C_{max}} = \frac{C}{C_{ox}} = \frac{C_d}{C_d + C_{ox}}$$

$$= \frac{4.3 \times 10^{-5}}{4.3 \times 10^{-5} + 3.45 \times 10^{-4}} = 0.11$$

Summary

Main symbols introduced

Symbol	Unit	Description
c	$C\,m^{-3}$	charge per unit volume
C_d	$F\,m^{-2}$	depletion region capacitance
C_{ox}	$F\,m^{-2}$	oxide capacitance
d	m	depletion region width
N_v	m^{-3}	effective density of states in the valence band
Q_d	$C\,m^{-2}$	depletion region charge
Q_f	$C\,m^{-2}$	fixed oxide charge
Q_i	$C\,m^{-2}$	interface trapped charge
Q_m	$C\,m^{-2}$	mobile ionic charge
Q_{ms}	$C\,m^{-2}$	charge due to work function difference
Q_n	$C\,m^{-2}$	inversion layer charge (p-type bulk)
Q_o	$C\,m^{-2}$	lumped oxide charge
Q_t	$C\,m^{-2}$	trapped oxide charge
t_{ox}	m	oxide thickness
V_F	eV	Fermi and intrinsic Fermi level separation, in voltage
V_{fb}	V	flat-band voltage
V_G	V	gate potential with respect to the semiconductor bulk
V_{TH}	V	threshold voltage (ideal)
$V_{TH'}$	V	threshold voltage (non-ideal)
V_{ox}	V	voltage drop across oxide
ϵ	$F\,m^{-1}$	permittivity of material ($= \epsilon_0 \epsilon_r$)
ψ_s	V	surface potential (oxide–semiconductor boundary)
ϕ_{ms}	V	metal–semiconductor work function difference

Main equations

$$qV_F = E_{Fi} - E_F \tag{14.1}$$

$$|Q_d| = (4\epsilon_s q N_A V_F)^{1/2} \quad \text{(at threshold)} \tag{14.16}$$

$$V_F = \frac{kT}{q} \ln \frac{N_A}{n_i} \quad (p\text{-type}) \tag{14.18}$$

$$V_G = \psi_s + V_{ox} \tag{14.7}$$

$$\psi_s = 2V_F \quad \text{(at threshold)} \tag{14.14}$$

$$V_{TH} = \frac{Q_d}{C_{ox}} + 2V_F \quad \text{(threshold } Q_d \text{ value)} \tag{14.15}$$

$$V_{fb} = \phi_{ms} - \frac{Q_o}{C_{ox}} \tag{14.23}$$

$$V_{TH'} = V_{TH} + V_{fb} \tag{14.24}$$

Exercises

(Take $T = 300$ K and silicon dioxide relative permittivity $= 3.9$ if required)

1. Draw band diagrams for equilibrium, accumulation, depletion and inversion for an ideal MOS capacitor with n-type semiconductor.

2. Find the oxide capacitance and charge m^{-2} for a 95 nm thick SiO$_2$ layer with an oxide voltage of 3 V. Find the maximum charge storage on a 1 mm^2 MOS capacitor if the oxide breaks down at 3.10^8 V m^{-1}.

3. Find the depletion region thickness for an MOS capacitor in depletion with a surface potential of 0.2 V, doped $N_A = 8.10^{23}$ and $N_D = 2.10^{23}$ m^{-3}. Find the gate potential under these conditions for a 100 nm oxide. Will the inversion layer in this device consist of electrons or holes? What will the surface potential be at threshold?

4. Find the threshold voltage for an ideal silicon MOS capacitor with $N_A = 10^{22}$ m^{-3} and $t_{ox} = 50$ nm at 300 K. Find the shift in threshold voltage if there are 10^{15} electronic charges m^{-2} at the interface.

5. It is equally important *not* to form an inversion layer in some regions, although they are structured like an MOS capacitor. Suggest two ways to achieve this during device manufacture.

6. An ideal MOS capacitor is formed from the insulator of (2) and the semiconductor of (3). Find the threshold voltage, and the ratio of maximum to minimum capacitance in a C/V plot. Sketch the C/V plot for high frequency excitation.

7. In bias–temperature stressing, a C/V curve is taken and then retaken after applying a dc voltage for some time at elevated temperature. What will this test reveal?

8. An n-type MOS capacitor has a threshold voltage of -2.5 V and an oxide thickness of 75 nm. Find the charge density of holes, if possible, at gate voltages of $+1$, -2 and -4 V.

9. Find the current which must flow if the MOS capacitor in (8) is to be switched between gate voltages of -0.5 and -4.5 V in 1 μs, and the capacitor area is 0.1×0.1 mm.

10. Find the threshold voltage for a p-type silicon MOS capacitor with $N_A = 10^{23}$ m^{-3}, $t_{ox} = 80$ nm, and $\phi_{ms} = 0.9$ V. Find ϕ_{ms} and V_{TH} for an n-type capacitor with the same doping density and oxide thickness.

15 MOS transistors

15.1 Description

Metal–oxide–semiconductor field effect transistors (MOSFETs or MOSTs) are probably the commonest solid state device in the world today. At the time of writing single packages containing in excess of one million MOSFETs are not unusual, and there is little sign of the trend to high packing densities levelling out. In this chapter we will give a physical and electrical description of this phenomenally successful device and outline some of its major applications.

A cross-section and top view of a MOSFET is given in Figure 15.1. There are three electrical connections, plus a possible substrate connection, as for the JFET. The *source* and *drain* regions have the same doping type and have ohmic metal contacts. The gate contact is separated from the channel region by a thin insulating film, normally silicon dioxide. In nearly every case the channel doping is of opposite type to the source and drain.

The MOSFET consists of an MOS capacitor, with contacts for lateral current flow in any inversion layer that may be formed. The flow can be regulated or interrupted by applying a gate voltage above or below the threshold

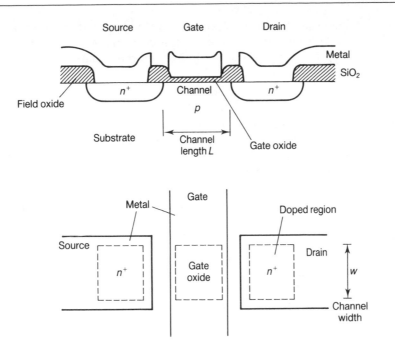

Figure 15.1 *n*-channel MOSFET cross-section and top view

voltage V_{TH}. If an inversion layer is not formed current does not flow between the source and drain, because either the source–channel diode or the channel–drain diode will be reverse biased.

Electrically there are four main styles of MOSFET, according to the doping of the channel and the polarity of the threshold voltage:

1. *n*-channel: the inversion layer contains electrons, and the channel is doped *p*-type.
2. *p*-channel: the inversion layer contains holes, so the channel is doped *n*-type.

The channel designation is therefore the same as the source and drain doping, and *opposite* to the doping in the channel. This causes confusion at times, so be careful when reading different texts. The other description is:

3. Depletion mode: a conducting channel exists without applying a gate voltage.
4. Enhancement mode: a gate voltage of the correct polarity and sufficient size is required to create a conducting channel.

A simple memory aid is to note that the depletiON mode device is normally ON (i.e. conducting from source to drain) before a gate voltage is applied. If the channel type and mode are both known a reasonably complete physical and electrical description can be deduced. For example, an n-channel, depletion mode device has:

1. n-type source and drain regions.
2. p-type semiconductor under the gate oxide.
3. A negative threshold voltage (conducting channel exists for $V_G = 0$).
4. An increase in source–drain current for $V_G > 0$.
5. Very little source–drain current for $V_G \ll 0$ ($V_G < V_{TH}$).

Figure 15.2 shows the circuit symbols for the four styles of MOSFET. There is no internationally agreed list of symbols, so students will encounter variants, such as enclosing the device in a circle. In each case a broken source–drain line indicates a normally off (enhancement mode) device, while a full line is for a normally on device. The arrow represents the pn diode formed by the channel substrate and the inversion layer (when formed). It points from p to n, i.e. towards the gate in an n-channel MOSFET.

Simplified construction steps for making a p-channel enhancement mode device are outlined in Figure 15.3:

1. Grow a thick field oxide on an n-type wafer. Metal tracks running over this will be less likely to cause unwanted inversion layers in between MOSFETs.
2. Mask (1). Define and etch windows for the source, gate and drain (Figure 15.3a).
3. Grow a thin gate oxide. The field oxide becomes slightly thicker (Figure 15.3b).
4. Mask (2). Define and clear windows for the source and drain regions, leaving the gate oxide intact.
5. Diffuse the source and drain p-type (Figure 15.3c).
6. Deposit metal.
7. Mask (3). Define and etch metal tracks (Figure 15.3d).

The gate oxide is grown separately because it is done in a slow dry oxidation to obtain the highest quality of interface. In this sequence we have assumed that field oxide spacers separate the gate metal from the source and drain contacts: more compact alternatives will be described later. The main point to notice is that only three photolithographic steps were required to make the device to this point, compared to four or five for a bipolar junction transistor. This has a major impact on yield, because each additional processing step can add more device-killing defects. In practice the possible yield improvement is traded for an increase in packing density.

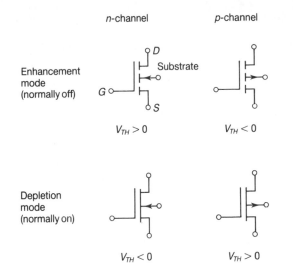

Enhancement
mode
(normally off)

Depletion
mode
(normally on)

Note:
Source and drain are often interchangeable. The arrow shows the polarity of the substrate – channel
pn diode when an inversion layer is formed.

Figure 15.2 MOSFET circuit symbols

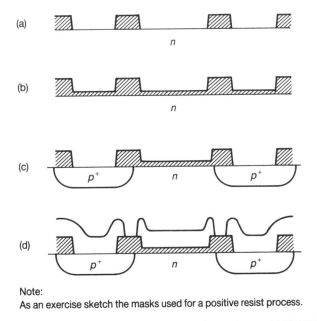

Note:
As an exercise sketch the masks used for a positive resist process.

Figure 15.3 Steps in the construction of a p-channel MOSFET

15.2 *I/V* characteristics

The circuit to be considered is given in Figure 15.4. The source is common to the input and output circuits. We are concerned with the drain–source current I_{DS} which flows in response to a drain–source voltage V_{DS} when a gate–source voltage V_{GS} is applied. For the *n*-channel device shown both V_{GS} and V_{DS} are positive in the normal region of operation; polarities of V_{GS}, V_{DS} and I_{DS} are reversed in a *p*-channel MOSFET.

So far we have only considered the existence or non-existence of a conducting channel between the source and drain. Now the nature of the transition will be explored, subject to the following simplifying assumptions:

1. Source and substrate are shorted together.
2. Carriers in the inverted channel have a *surface mobility* $\bar{\mu}_e$ which is about half the bulk value. They are also subject to velocity saturation at high fields.
3. A gradual channel approximation applies, so that only horizontal current flow need be considered.
4. V_{TH} is constant along the channel.
5. No subthreshold current flows. ($I_{DS} = 0$ when $V_{GS} < V_{TH}$).
6. The surface potential just under the gate oxide $V_s(y)$ varies with position y along the channel, from $V_s(0) = 0$ at the source to $V_s(L) = V_{DS}$ at the drain. (Choosing the $+y$ direction in this sense will make the source–drain current I_{SD} positive for conventional current flowing from source to drain. The drain–source current $I_{DS} = -I_{SD}$)

When $V_{GS} = V_{TH}$ the gate voltage supports Q_d, the depletion region charge, but little mobile inversion charge:

$$Q_d = -C_{ox}V_{TH} \text{ m}^{-2} \tag{15.1}$$

when $V_{GS} > V_{TH}$ the depletion charge remains almost static and the excess

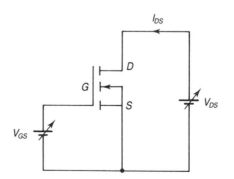

Figure 15.4 Voltage and current definitions

voltage supports an inversion charge $Q_n(y)$:

$$Q_n(y) = -C_{ox}(V_{GS} - V_{TH} - V_s(y))\,\mathrm{C\,m}^{-2} \tag{15.2}$$

$(V_{GS} - V_{TH})$ is the voltage above threshold in an MOS capacitor. $V_{GS} - V_{TH} - V_s(y)$ is the *local* voltage excess at an arbitrary point along the channel, and must remain positive for an inversion charge to exist.

The conventional source–drain current I_{SD} depends on the magnitude of $Q_n(y)$ and on how fast it is moving. The drift velocity in the $+y$ direction in the inversion layer, for low lateral fields, is:

$$v_{drift} = -\bar{\mu}_e \frac{(-\mathrm{d}V_s)}{\mathrm{d}y}\,\mathrm{m\,s}^{-1} \quad \text{(negative charge)} \tag{15.3}$$

and the current is:

$$I_{SD} = W\bar{\mu}_e Q_n \frac{\mathrm{d}V_s}{\mathrm{d}y}\,\mathrm{A} \tag{15.4}$$

where W is the channel width perpendicular to current flow (Figure 15.1). (Check the units of the last equation.)

Substituting for $Q_n(y)$, we obtain an equation only in $V_s(y)$, together with circuit and physical parameters:

$$I_{SD} = -W\bar{\mu}_e C_{ox}(V_{GS} - V_{TH} - V_s(y))\frac{\mathrm{d}V_s(y)}{\mathrm{d}y} \tag{15.5}$$

The variables V_s and y can be separated, and the resulting equation integrated along the channel, as in the JFET analysis:

$$\int_0^L I_{SD}\,\mathrm{d}y = -W\bar{\mu}_e C_{ox}\int_{V_s=0}^{V_s=V_{DS}} (V_{GS} - V_{TH} - V_s)\,\mathrm{d}V_s \tag{15.6}$$

I_{SD} is constant along the channel, so:

$$I_{SD} = \frac{-W\bar{\mu}_e C_{ox}}{L}[(V_{GS} - V_{TH})V_{DS} - V_{DS}{}^2/2] \tag{15.7}$$

This is valid provided an inversion layer exists at all points along the channel. Inversion is weakest at the drain end, where the drain potential is tending to counter the influence of the gate. The range is therefore:

$$V_{GS} - V_{TH} \geqslant V_{DS} \quad (n\text{-channel}) \tag{15.8}$$

and I_{SD} turns out negative, i.e. conventional current flows from drain to source when V_{DS} is positive, as common sense suggests. The graphical form of equation (15.7) is given in terms of a positive drain–source current in Figure 15.5.

Pinch off occurs in MOSFETs when the theory above predicts that $Q_n(L)$ falls to zero, as represented diagrammatically in Figure 15.6. The model breaks down at this point because:

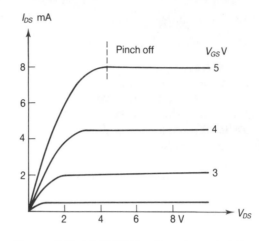

Figure 15.5 MOSFET I_{DS}/V_{DS} characteristic

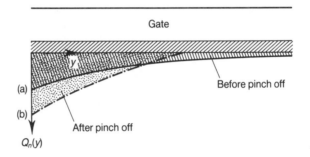

Figure 15.6 Inversion charge Q_n along the channel as predicted by simple theory

1. Carriers move at a constant saturation velocity in the pinch off region and cannot accelerate without limit.
2. The gradual channel approximation fails at pinch off.

In practice the current I_{DS} rises with V_{DS} until pinch off, as given by equation (15.7), and then nearly levels out. The saturation current $I_{DS(\text{sat})}$ is approximately found by putting $V_{DS} = V_{GS} - V_{TH}$ into equation (15.7):

$$I_{DS(sat)} = \frac{W \bar{\mu}_e C_{ox}}{2L}(V_{GS} - V_{TH})^2 \text{ A} \tag{15.9}$$

The family of I_{DS}/V_{DS} curves are not equally spaced, but follow a square law. An important circuit parameter in the saturation region is the mutual conductance or transconductance g_m, which is the variation of the output current I_{DS} with the input voltage V_{GS}:

$$g_m = \frac{\partial I_{DSsat}}{\partial V_{GS}} = \frac{W \bar{\mu}_e C_{ox}}{L}(V_{GS} - V_{TH}) \, \Omega^{-1} \tag{15.10}$$

The dependence of g_m on V_{GS} underlines the non-linearity of the MOSFET. This equation also brings out the importance of the size ratio W/L. High gain requires a short source–drain channel length L and a wide device perpendicular to the main current flow. These requirements coincide with those for a minimum channel resistance before saturation.

EXAMPLE

◼ Find the current I_{DS} flowing in a p-channel MOSFET for which:

$W = 50 \, \mu m$ $L = 5 \, \mu m$ $\bar{\mu}_e = 0.02 \, \text{m}^2 \, \text{V}^{-1} \, \text{s}^{-1}$

$t_{ox} = 0.1 \, \mu m$ $V_{TH} = -2V$ $V_{DS} = -10 \, \text{V}$ $V_{GS} = -8 \, \text{V}$

☐ First determine the region of operation. V_{GS} is more negative than V_{TH}, attracting holes to the interface. This p-channel MOSFET is inverted and current flows. The equivalent relation to equation (15.8) for a p-channel device is:

$$V_{GS} - V_{TH} \leqslant V_{DS} \quad (p\text{-channel}) \tag{15.11}$$

to avoid pinch off. (The quantity $V_{GS} - V_{TH}$ must be lower than the surface potential V_{DS} at the drain in order to support a positive inversion charge.) Here $V_{GS} - V_{TH} = -6 \, \text{V}$ while $V_{DS} = -10 \, \text{V}$, so the transistor is in saturation. Equation (15.9) then gives I_{DS}, with a sign change for p-channel.

$$I_{DS(sat)} = \frac{-5 \times 10^{-5} \times 0.02 \times 3.9 \times 8.85 \times 10^{-12}}{2 \times 5 \times 10^{-6} \times 10^{-7}}(-8 + 2)^2$$

$$= -6.2 \, \text{mA}$$

15.3 Small-signal and switching

15.3.1 Small signals

A simple small-signal equivalent circuit for a MOSFET is given in Figure 15.7. The source is taken as common to the input and output circuits, the input is a small ac voltage v_{gs} applied to the gate, and the output is a small-signal voltage v_{ds} at the drain. It is necessary to stipulate *small* ac signals because the circuit element values depend on the voltage across them.

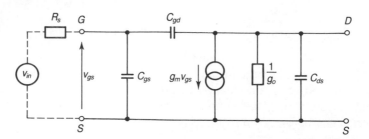

Figure 15.7 Small-signal equivalent circuit, common source configuration

The dc input impedance is essentially infinite because of the insulating gate oxide, but the gate–source and gate–drain capacitances, C_{gs} and C_{gd}, are important to ac input signals. The output characteristic is dominated by the mutual conductance, i.e., the drain–source current which flows in response to the gate–source voltage. This current is split between any load impedance, the output resistance $1/g_o$ and the drain–source capacitance C_{ds}. g_o is the output admittance, which is the slope on the I_{DS}/V_{DS} graph in saturation, at constant gate voltage:

$$g_o = \left. \frac{\partial I_{DS}}{\partial V_{DS}} \right|_{V_{GS}=\text{constant}} \Omega^{-1} \tag{15.12}$$

The ideal dc theory made g_o zero, but I_{DS} increases with V_{DS} in real devices, mainly due to channel shortening. Typical values for a discrete device are:

$C_{gs} = 4\ \text{pF}$

$C_{gd} = 1\ \text{pF}$

$C_{ds} = 2\ \text{pF}$

$g_o = 2 \times 10^{-5}\ \Omega^{-1}$

$g_m = 4 \times 10^{-3}\ \Omega^{-1}$

The output resistance is $1/g_o = 50\ \text{k}\Omega$. The capacitances can often be neglected in low frequency analysis, where $1/\omega C$ is large compared to the load resistance.

EXAMPLE

■ Find the small-signal voltage developed across a $1\ \text{k}\Omega$ load resistor with $C_{gs} = 5\ \text{pF}$, $C_{gd} = C_{ds} = 1\ \text{pF}$, $g_o = 4 \times 10^{-5}\ \Omega^{-1}$, $g_m = 5 \times 10^{-3}\ \Omega^{-1}$, if $v_{gs} = 100\ \text{mV}$ at a frequency of $100\ \text{kHz}$.

□ The impedance of C_{gd} and C_{ds} at $100\ \text{kHz}$ is:

$$X_C = 1/2\pi f C \; \Omega \quad \text{reactive} \tag{15.13}$$

$$= 1/(2\pi \times 10^5 \times 10^{-12})$$

$$= 1.59 \; \text{M}\Omega$$

This is large compared with the load resistance and with $1/g_o$ ($= 25 \; \text{k}\Omega$), so the currents in these capacitors can be neglected to a first approximation. The small-signal drain–source current is:

$$i_{ds} = g_m v_{ds} = 5 \times 10^{-3} \times 0.1$$

$$= 5 \times 10^{-4} \; \text{A}$$

The current flows through a parallel combination of $25 \; \text{k}\Omega$ and the load resistance $1 \; \text{k}\Omega$. The current through $1 \; \text{k}\Omega$ is therefore:

$$i_{Load} = \frac{25}{25 + 1} \times 5 \times 10^{-4}$$

$$= 0.48 \; \text{mA}$$

The voltage developed is:

$$v_{Load} = i_{Load} R_{Load} = 0.48 \; \text{V}$$

15.3.2 High frequency

The ac equivalent circuit can also be used to derive the high-frequency behaviour of a MOSFET. The gate-source capacitance C_{gs} is the dominant element here, because it tends to short out the input at high frequencies. This does not matter if the transistor is driven by a perfect voltage source which can supply all the reactive current, but real circuits will have a source resistance R_s as in Figure 15.7. The current drawn by C_{gs} causes a voltage drop across R_s which reduces v_{gs} at the device and decreases the gain above the break-point frequency $1/R_s C_{gs}$.

Although the MOSFET is a majority carrier device, free from bipolar minority carrier storage effects, the mobile charges in the channel still requires a finite time to pass from source to drain. For a channel length L and a charge saturation velocity v_{sat} the maximum response frequency is of order v_{sat}/L Hz.

Integrated circuits which use doped polysilicon as the gate conductor tend to be limited by the RC delay of the gate and the conductor–substrate capacitance. Devices which use metal for the gate interconnection tend to be limited by the carrier transit time within the device.

EXAMPLE

■ A MOS integrated circuit has gate lengths of $1 \, \mu m$, gate width of $5 \, \mu m$, 30 nm oxide thickness and polysilicon interconnect tracks $2 \, \mu m$ wide with a sheet resistance of $20 \, \Omega/\text{square}$. The saturation velocity is $10^5 \, \text{m s}^{-1}$. Find the maximum frequency of operation if the gate interconnections are $50 \, \mu m$ long.

☐ The gate capacitance is assumed to be dominant:

$$C_{gs} = \frac{(\text{gate area})\epsilon}{\text{oxide thickness}} = \frac{10^{-6} \times 5 \times 10^{-6} \times 3.9 \times 8.85 \times 10^{12}}{30 \times 10^{-9}}$$

$$= 5.7 \times 10^{-14} \, \text{F}$$

The source resistance of the track is found by multiplying the sheet resistance by length of the track in $2 \, \mu m$ wide squares:

$$R_s = 20 \times 50/2 = 500 \, \Omega$$

The maximum frequency from this cause is:

$$1/R_s C_{gs} = 3.5 \times 10^{10} \, \text{Hz}$$

This contrasts with the transit-time-limited frequency:

$$v_{sat}/L = 10^5/10^{-6}$$

$$= 10^{11} \, \text{Hz}$$

The lower frequency limit of 35 GHz due to the RC time constant applies.

15.3.3 Switching

Most MOSFETs are used in digital applications, where the device is required to be either on and conducting or off and non-conducting. As in bipolar switching circuits the MOSFET should mimic as far as possible the action of a perfect mechanical switch:

1. Low on resistance, R_{ON}.
2. High off resistance, R_{OFF}.
3. Short switching time.

A basic MOSFET switching circuit is shown in Figure 15.8. The input voltage V_{in} is applied to the gate of an n-channel enhancement mode MOSFET. V_{in} can be either V_L ($< V_{TH}$) or V_H ($\gg V_{TH}$) in the low or high digital conditions. The circuit voltage V_{DD} is shared between the MOSFET and the

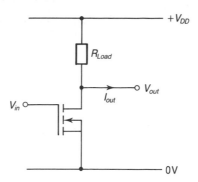

Note:

V_{in} can be low ($V_L < V_{TH}$) or high ($V_H \gg V_{TH}$). R_{Load} may be a resistor or another MOSFET.

Figure 15.8 MOSFET inversion circuit

load, with the output voltage (and current) taken at their junction. A well-defined digital output requires that:

1.   Load impedance $R_{Load} \gg R_{ON}$.
2. $R_{Load} \ll R_{OFF}$.
3.   For an output current I_{out}, $R_{Load}I_{out} \ll V_{DD}$.

These conditions imply that $V_{out} \simeq V_{DD}$ when the MOSFET is off and $V_{out} \simeq 0$ when it is turned on, making the circuit an inverter. Too high an output current would cause V_{out} to fall to an undefined intermediate level, because of the $R_{Load}I_{out}$ voltage drop across the load. The load in a final power stage would be in the position of R_{Load}. More likely, the inverter would be an intermediate stage with other, external load impedances connected to it. The term load is not well defined in this application. This poses few problems if the external load(s) are other MOSFET gates which take a negligible input current except when switching.

The series load could be a simple resistance in a circuit composed of a few discrete devices, but not in an integrated circuit where the total power dissipated must be held to a minimum. This is brought out in the following example.

EXAMPLE

■   Consider an integrated circuit which contains 10^5 MOSFET inverters each with a load resistance R_{Load}. It runs from a 5 V supply and the package can dissipate a maximum of 1 W. The load resistors are formed from a meander of polysilicon with a $1\,\text{k}\Omega/\text{square}$ sheet resistance and a $5\,\mu\text{m}$ minimum feature size. Find the minimum value of R_{Load} and the chip area required for the load resistors, assuming half the stages are on at any one time.

☐ One watt maximum power with a 5 V supply gives a maximum load current of 0.2 A. Therefore:

$$\text{Max load current/inverter} = \frac{0.2}{0.5 \times 10^5} = 4\,\mu\text{A}$$

If the full supply voltage is dropped across R_{Load}, the minimum value for R_{Load} is:

$$R_{Load} = \frac{5}{4 \times 10^{-6}} = 1.25\,\text{M}\Omega$$

Using a $1\,\text{k}\Omega/\text{square}$ sheet resistance (a rather high figure) this gives meanders 1250 squares long, or 6.25 nm long with a $5\,\mu\text{m}$ square size. The tracks for 10^5 such resistors would occupy an area:

$$A = 1250 \times 5 \times 10^{-6} \times 5 \times 10^{-6} \times 10^5$$
$$= 3.1 \times 10^{-3}\,\text{m}^2$$

with an equivalent area of spacing between tracks, giving a total load resistor area of $6.2 \times 10^{-3}\,\text{m}^2$, which fills a square 79×79 mm. This is far larger than the largest die fabricated today.

In practice load impedances are formed from other MOSFETs, the two most popular being:

1. An enhancement mode MOSFET with gate and drain connected.
2. A depletion mode MOSFET with gate and source connected.

The I/V characteristics of these MOS loads are shown in Figure 15.9. They can be deduced from the MOSFET characteristics, putting $V_{GS} = V_{DS}$ and $V_{GS} = 0$ in the two cases. Both give a rise in current for an increase in voltage, the first slower than linear and the second faster than linear at low load currents. They occupy a similar area as the switching transistor, about $20 \times 20\,\mu\text{m}$ for a $5\,\mu\text{m}$ gate length. The depletion mode load gives faster switching, because it rapidly provides more current when switching begins, which charges the external load capacitance faster.

Although MOS loads can be made physically smaller than resistors, the circuit still draws current about half the time in a digital IC. The solution is to make both components active switches, but with a complementary action. The complementary MOS (CMOS) inverter is shown in Figure 15.10. It consists of

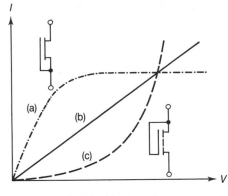

Note:
(a) Depletion mode MOSFET with gate and source connected.
(b) Resistor
(c) Enhancement mode MOSFET with gate and drain connected.

Figure 15.9 *I/V* characteristics of loads

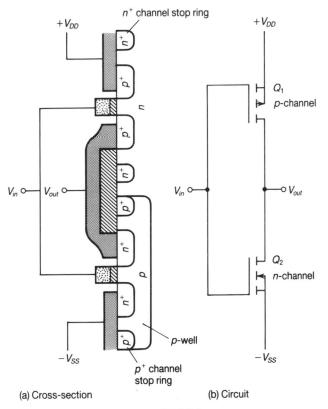

(a) Cross-section   (b) Circuit

Figure 15.10 CMOS inverter

an n-channel and a p-channel MOSFET, with their gates connected together. When made on the same substrate one MOSFET, in this case Q1 is formed directly on the n-type wafer. The other is fabricated in a well or tub of opposite doping. The most positive circuit potential is connected to the n-type substrate and the most negative to the p-wells, to provide junction isolation between the two types of MOSFET. Additional rings of highly doped material surround each transistor to inhibit parasitic MOSFETs from turning on. These are called channel stops.

Virtually no static current is drawn because only one transistor is conducting at any one time. Power is dissipated principally through the charging and discharging of the output capacitance C_o at a switching frequency f Hz:

$$P = C_o f V_{DD}^2 \ \text{W} \tag{15.14}$$

Applications with low switching frequencies can have extremely low power dissipations, making CMOS especially suitable for battery-driven circuits such as digital watches.

15.4 Non-ideal behaviour

Real MOSFETs depart from the ideal characteristics described in the previous sections, particularly at the extremes of current and voltage. The main physical causes of non-ideal behaviour are discussed in this section, together with their effects on the I/V characteristics.

15.4.1 Subthreshold current

The transition between the conducting and non-conducting states has been assumed to be sharp at the threshold voltage. However, the band diagram of the MOS capacitor made it clear that inversion is a progressive phenomenon. This becomes more important in scaled down, low power circuits, where the threshold voltage is reduced. Figure 15.11 shows a detail of the I_{DS}/V_{GS} characteristic for a large MOSFET and for a scaled down device. The subthreshold current does not scale, giving the smaller device a poorer digital noise immunity.

15.4.2 Channel avalanche

A charge multiplication avalanche can occur in the pinch off region at high V_{DS}. In an n-channel device the extra electrons join the drain current while the holes flow to the substrate. The overall effect is an increase in I_{DS} above that predicted by the simple model. In a short channel device, with a gate length

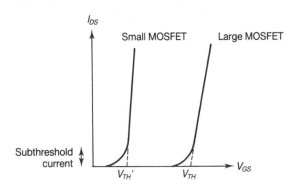

Note:
The current for $V_{GS} \leqslant V_{TH}$ is nearly the same in large and in scaled down devices.

Figure 15.11 Detail of I_{DS}/V_{GS} characteristic

below 3 μm, the hole current to the substrate can initiate a strange series of events:

1. A large fraction of holes flow to the nearby source.
2. The (hole current) × (substrate resistance) voltage drop forward biases the source–substrate diode.
3. The source injects electrons into the substrate, which flow on to the drain.

The MOSFET then operates in parallel with a parasitic bipolar junction transistor, formed by the source, substrate and drain.

15.4.3 Oxide charging

Electrons in the pinch off region can possess a drift velocity comparable to their thermal velocity. Some of these so-called hot electrons are accelerated towards the oxide by the pinch off field and gain sufficient energy to surmount the Si/SiO$_2$ conduction band potential barrier (3.2 eV). Some of these charges will be trapped within the oxide and cause a change in threshold voltage. This effect is undesirable in MOSFET circuits, but is the basis of several memory cells.

15.4.4 Punchthrough

The depletion region around the $n+$ drain region in an n-channel MOSFET extends some way into the substrate and towards the source (Figure 15.12). When the gate length is less than 0.5 μm it is possible for this depletion region

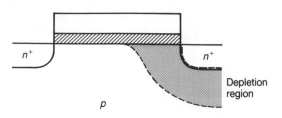

Figure 15.12 When the depletion region extends to the source the device has punched through

to reach as far as the source. Electrons in the $n+$ source then see a high lateral field pulling them through the bulk of the semiconductor, as well as through the surface inversion channel. This is the punchthrough condition for a MOSFET. The I_D/V_{DS} characteristic no longer displays a saturation.

EXAMPLE

■ Estimate the punchthrough voltage for a MOSFET with the following parameters: substrate doping $N_A = 10^{22}$ m^{-3}; source and drain doping $N_D = 10^{25}$ m^{-3}; channel length 1 μm.

☐ At punchthrough the full drain–source voltage V_{DS} is dropped across a depletion region which extends along the channel length. The asymmetry of doping means that a single-sided diode formulation can be used, which with a one dimensional approximation gives a depletion region width:

$$d_p = \left(\frac{2\epsilon V_{DS}}{qN_A}\right)^{1/2} \tag{15.15}$$

or:

$$V_{DS} = \frac{qN_Ad_p{}^2}{2\epsilon}$$

$$= \frac{1.6 \times 10^{-19} \times 10^{22} \times 10^{-6} \times 10^{-6}}{2 \times 11.9 \times 8.85 \times 10^{-12}}$$

$$= 7.6 \text{ V}$$

15.4.5 Oxide conduction

If in an MOS capacitor ϵ_{ox} and $\mathscr{E}_{ox}$ are the permittivity and electric field for SiO$_2$, and ϵ_s and $\mathscr{E}_s$ are the values for semiconductor, they are related at the interface by:

$$\epsilon_{ox}\mathscr{E}_{ox} = \epsilon_s\mathscr{E}_s \tag{15.16}$$

The peak field before avalanche in silicon is about $3 \times 10^7 \text{ V m}^{-1}$, corresponding to an oxide field of $9 \times 10^7 \text{ V m}^{-1}$. This is achieved with only 9 V on the gate of a $0.1 \ \mu\text{m}$ thick oxide. Even at this field simple electron or hole conduction through SiO_2 is extremely small, of order 10^{-7} A m^{-2}. This is not the case for silicon nitride or Al_2O_3 films, which are used in more advanced gate structures, where the current density at this field is of order 10^{-3} A m^{-2}.

If the thickness of the gate dielectric is only 10–20 nm, quantum-mechanical tunnelling begins to produce a significant leakage current, even in silicon dioxide. Mobile charges tunnel through the barrier without troubling to gain enough energy to reach the conduction band. Because the uncertainty principle forbids both the speed of an electron and its position to be known precisely, an electron localized to the very thin insulator behaves as if its energy were temporarily much higher, allowing it to appear on the other side.

Breakdown, involving damage to the insulator, occurs at higher field strengths than the semiconductor substrate can sustain for long periods. For SiO_2 the breakdown field is 9.10^8 V m^{-1} for thicknesses above $0.1 \ \mu\text{m}$, and rises to 3.10^9 V m^{-1} in very thin films. These fields are easily reached in transient events, necessitating the use of input protection circuitry in MOSFET integrated circuits. These typically take the form of diodes which break down quickly and non-destructively, diverting the charge to ground and dissipating the pulse energy before a damaging field level is reached. Precautions to avoid the build up of static charges are taken throughout the electronics industry, especially when handling these devices.

EXAMPLE

■ Suppose a worker has a capacitance to ground of 10 pF, and is handling unprotected MOSFETs with $0.1 \ \mu\text{m}$ gate oxides and gate areas of $10 \times 10 \ \mu\text{m}$. What static potential on the worker would cause the destruction of a gate?

□ The circuit conditions are those in Figure 15.13. The initial charge is $Q_0 = V_0C_1$. This is shared between the capacitances C_1 and C_2 in parallel when contact is made. In equilibrium the same voltage V_2 exists across the worker and the MOS capacitor, assuming they share a common ground. Charge conservation gives:

$$Q_0 = Q_1 + Q_2 \tag{15.17}$$

and:

$$V_2 = Q_1/C_1 = Q_2/C_2 \tag{15.18}$$

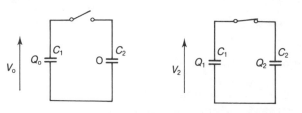

Figure 15.13 Circuit formed by MOSFET and worker

Hence:

$$V_0 = \frac{C_1 + C_2}{C_1} V_2 \qquad (15.19)$$

Here $C_1 = 10$ pF and:

$$C_2 = \frac{A\epsilon}{t_{ox}}$$

$$= \frac{10^{-10} \times 3.9 \times 8.85 \times 10^{-12}}{10^{-7}}$$

$$= 3.5 \times 10^{-14} \text{ F or } 0.035 \text{ pF}$$

For a breakdown field of 9.10^8 V m^{-1}, $V_2 = 90$ V at breakdown and $V_0 = 90.3$ V. Although this would be felt as an electric shock from a supply, static potentials of several kV are routinely measurable on unprotected personnel. This presents a severe hazard to MOSFET devices.

15.5 MOSFET families

The range of MOSFET devices is very broad, from the smallest integrated transistor to discrete devices capable of switching hundreds of amps. The names applied to the styles proliferate even faster as minor variants are given radically different names, probably in the hope of immortalizing the inventors. In this section we will outline the main functional families and technological improvements to the basic MOSFET structure. Memory devices are discussed in a separate section.

15.5.1 Self-aligned structures

The gate structure in Figure 15.14a performs well in integrated circuits constructed with 5–10 micron features, operating at up to 10–50 MHz. Improved

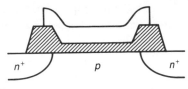

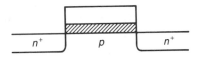

(a) Conventional MOSFET

(b) Self-aligned MOSFET

Figure 15.14 Gate structure

high frequency characteristics require

1. A shorter channel, to reduce the transit time.
2. Less overlap between gate metal and the source and drain doping, to reduce C_{gs} and C_{gd}.

This is achieved in the structure in Figure 15.14b by automatically aligning the source and drain doping with the edge of the gate. Aluminium can no longer be used as the gate metal, because it cannot withstand either a diffusion doping step or a dopant anneal step following ion implantation. Tungsten was briefly considered around 1970, but was difficult to deposit. The successful breakthrough was to use doped polycrystalline silicon, deposited by LPCVD. The sheet resistance can be as low as 10 ohms/square, making it suitable for short interconnection paths. It can be shrouded with a deposited dielectric, such as SiO_2 by a PECVD technique, allowing a second layer of metallic conductor paths above. As device geometries have shrunk, polysilicon is being supplemented with a refractory metal disilicide layer, such as WSi_2 or $MoSi_2$, to reduce the sheet resistance. These polycide layers form part of the salicide process (self-aligned silicide), which will eventually be replaced by a purely metallic gate again.

Ion implantation is the second key to a self-aligned structure, together with a high temperature compatible gate metal. Diffusion doping spreads laterally nearly as much as it does vertically but ion implantation is more nearly vertical. A little lateral spreading does occur due to the scattering of the implant beam and the diffusion during the implant activation anneal.

Polysilicon gates offer a further advantage, because the work function difference ϕ_{ms} between gate metal and semiconductor substrate can be minimized in this way. The threshold voltage is also reduced as compared to an aluminium gate. The stability of the polysilicon/SiO_2 interface is also attractive so that future gate metallization schemes may well include polysilicon as the first layer in a composite structure.

15.5.2 Power MOSFETs

At the other end of the spectrum are discrete and integratable devices capable of handling tens of amps when on and holding off hundreds of volts when off.

A very low on resistance is required to minimize the I^2R power dissipation within the device, which in turn calls for a very short channel length. Some ingenious solutions are in production which are not limited by photolithography to define the channel length.

Just as the bipolar transistor base width can be fixed by diffusion profiles, the DMOS transistor shown in Figure 15.15 achieves a 1–2 micron effective channel length by diffusing first acceptor boron then donor phosphorus in the source window. Boron has a higher diffusion coefficient than phosphorus, so a narrow band of p-type material forms ahead of the spreading source diffusion. This double diffused DMOS structure allows both source and drain to appear on the same side of the wafer. This gives the potential for integrated logic and power control to appear side by side.

Other power MOSFETs rely on a vertical current flow through the wafer, but use the same principle. Examples are shown in Figures 15.16 and 15.17. The HEXFET structure consists of an array of hexagons when viewed from above. The buried polysilicon gates trace the outlines of the hexagons while the source areas cover the main surface areas. Currents flow laterally under the polysilicon gates and swing downwards to the drain contact on the other side of the substrate.

The VMOS device relies on liquid etchants which attack $\langle 100 \rangle$ crystal surfaces faster than $\langle 111 \rangle$ surfaces, yielding a v-shaped groove. The notch penetrates a narrow p-layer formed by double diffusion, and subsequent oxidation and metallization creates the MOS capacitor there. The drain contact is again on the bottom of the wafer. Features of both HEXFET and VMOS architectures are:

1. Individual devices are connected in parallel to increase the maximum current capability.
2. MOS devices do not inject minority carriers and can be used at higher frequencies than comparable bipolar devices – typically above 1 MHz.
3. Devices are built in a lightly doped epitaxial layer, to avoid punching through the narrow channel. The diffused channel region has a higher doping density than the underlying epitaxy, so the depletion region tends to extend downwards. Operation is possible at up to 200 V for VMOS and 600 V (and rising) for HEXFETs.

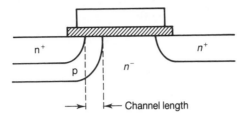

Figure 15.15 Double diffused MOS: DMOS

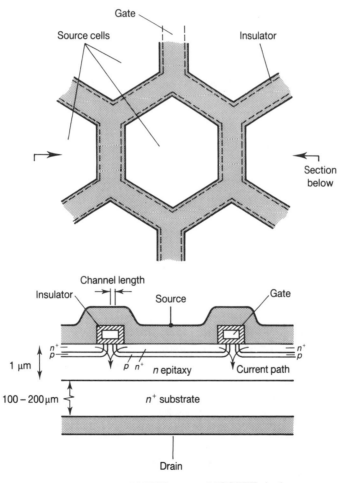

Figure 15.16 HEXFET power MOSFET design

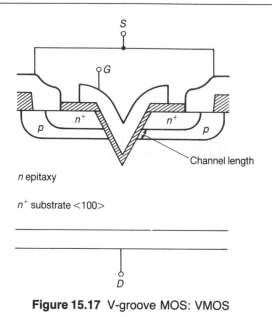

n epitaxy

n^+ substrate <100>

Figure 15.17 V-groove MOS: VMOS

4. The channel width Z measured around the perimeter of a v-groove or hexagon is very large compared to the channel length L. A large Z/L ratio increases g_m and raises I_{DS}.

15.5.3 Silicon on insulator (SOI) structures

Integrated MOSFET circuits are self-isolating because the source and drain regions form reverse biased *pn* diodes with the substrate. However, these isolation diodes add to the capacitance which must be charged when switching a MOSFET. It is almost possible to eliminate these parasitic capacitors by making the MOSFETs in a thin, single-crystal layer of silicon deposited on an insulating substrate.

Silicon can be deposited epitaxially on a variety of crystalline insulators, particularly sapphire (Al_2O_3) and spinel ($MgAl_2O_4$). It is also possible to recrystallize polysilicon deposited on amorphous insulators, such as SiO_2 and silicon nitride, by heating the film rapidly. (Crystalline insulator structures are sometimes termed SOS or silicon on sapphire, while the designation SOI is reserved for other insulators.) The technology is far from trivial, limiting the devices mainly to military and space markets, where a high immunity to radiation damage is demanded. It is beginning to compete with conventional CMOS because *p*-type and *n*-type regions can be formed separately without the need for building wells.

SOI devices are limited by a high defect density and a low minority carrier lifetime (below 10 ns) arising from mismatches between the silicon film and the substrate. The film is also under compression because sapphire has a thermal expansion coefficient almost twice that of silicon. This reduces carrier mobilities below the bulk values. Deposition at lower temperatures may assist in this problem.

15.6 Memory structures

The power of integrated logic circuits to process information is greatly enhanced by the ability to store digital data on silicon chips. Up to 16 megabits $(16 \times 1,024 \times 1,024$ bits) of data can be stored on a single chip at present (1990), enough to hold two hundred pages of text. (8 digital *bits* make one *byte*. One byte can store one alpha-numeric character. Memories are specified in both bits and bytes, with the conventions 2^{10} bit $= 1,024 = 1$ kilobit, 2^{13} bits $= 8 \times 1,024 = 1$ kilobyte. A 64k RAM could be $64 \times 1,024$ bits *or* bytes, so be careful.) Semiconductor memories are relatively cheap, are less environmentally sensitive than magnetic media and have a considerably shorter access time.

The types of memory device are summarized in Table 15.1. They divide broadly into read only memory (ROM), which holds fixed data indefinitely, and read/write or random access memory (RAM), where each bit can be altered at will. (No one has successfully marketed the write only memory or WOM yet!) RAMs are divided into:

1. *Static RAM*: each bit is held in a digital latch structure of 4–6 transistors. Data is present until the power is removed.

Table 15.1 Memory devices

Function	Type	Data storage time	Read access time (ns)	Availability (bits) 1990
Read/write (Random access memory)	Dynamic RAM – DRAM	ms unless refreshed	100	16K–4M
	Static RAM – SRAM	While powered	30–100	16K–64M
Read (Read only memory)	Erasable PROM – EPROM	Until erased	200	16K–2M
	Electrically erasable – E²PROM	Until altered ⎫	250	1K–256K
	Electrically alterable – EAROM	⎬ even without power		
	Programmable – PROM	Permanent ⎭		

2. *Dynamic RAM*: each bit is stored as a charge on an integrated capacitor, with 1–4 additional transistors per bit. The charge must be refreshed every few milliseconds by auxiliary circuitry, which is shared by several bit cells. Again data is lost if the power fails.

The need for continuous power has had a marked effect on the battery industry. RAMs, especially SRAMs are routinely backed up by local power cells which can extend the memory for hours or even weeks beyond external power failure. Some SRAMs even contain a back up battery within the chip package.

The alternative ROMs were originally programmed during manufacture by the omission or inclusion of a MOSFET load at the intersection of row and column address lines. Virtually all devices are now *programmable* in the workplace and are called PROMs. The main varieties are:-

1. *Programmable ROM – PROM*: the memory comprises an array of fusible metal links, which can be selected and blown individually by an excess current. Once blown a link cannot be remade.

2. *Erasable PROM – EPROM*: data is stored as a charge on the isolated polysilicon gates of the floating gate avalanche MOS device (FAMOS); see Figure 15.18. The bit is set by avalanche charge injection when an excess voltage V_{DS} is applied. The package contains a window so that all bits can be reset by irradiating it with ultra-violet light. This energises the trapped charges to the insulator conduction band.

3. *Electrically alterable ROM – EAROM*.

4. *Electrically erasable PROM – E^2PROM*: the same memory exists under two names. Like the EPROM, charge is stored on a floating polysilicon gate. Charges can be introduced and removed via a second gate, hence the name stacked gate avalanche MOS or SAMOS (Figure 15.19). The metal–nitride–oxide–semiconductor (MNOS) transistor is an alternative basis for the E^2PROM. Charge can be injected electrically to the nitride/oxide interface, where it remains trapped for very long periods.

The main electrical characteristics of these devices are summarized in Table 15.1. There is a trade off between speed and power dissipation which is fundamental because current flows to charge and discharge the storage capacitors. The faster data is switched, the higher the mean current and the greater the power dissipation.

Memory fabrication places the highest demands on semiconductor technology. A high-density memory chip less than 10 mm square is written with enough detail to record a street map of London or New York down to the smallest cul-de-sac. Perfect copies must be made from error-free masters overlaid up to

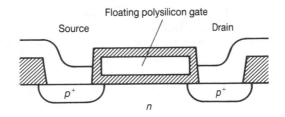

Note:
There is no external connection to the polysilicon gate.

Figure 15.18 Floating-gate avalanche MOS: FAMOS

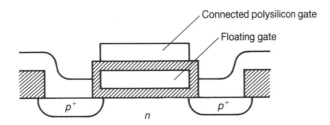

Figure 15.19 Stacked-gate avalanche injection MOS: SAMOS

ten layers deep, like a multilayer map showing water, power, subways and sewage. Even when it is finished errors can creep in due to alpha-emitting impurities in the packaging material, which create extra charges and can change the state of a memory cell. Rather than aiming for absurd perfection designs now tolerate a few bad cells by routing the data elsewhere, by using error correction codes or by allowing the remaining good memory to be used. Still higher packing densities are achievable even with present technology: the limiting factor is more the high cost of developing a new generation of devices.

Postscript

This book is most probably being read during professional training at college. How much of this knowledge will still be taught when the reader is in mid-career? Despite the rapid pace of technological change, a good part of this text is likely to remain relevant. It is intended to be a core text, dealing with basic physical principles and providing tools for more complex tasks. Junction diodes, doping and MOSFET-style transistors will persist for some decades yet. Although gallium arsenide and photonic devices will eat into the silicon market, superconducting Josephon devices and other exotics will remain at the periphery.

As to the kind of world it will be (provided wars and rumours of wars are postponed) electronics offers immense potential for change. The direction of change is open to choice: the same technology can provide both a more open, safer, culture through such factors as access to information, a wider selection of entertainment, and collision-avoidance systems for cars, and a more menacing future of unchallengeable databases, close monitoring of individuals and smart weapons. There is a rare opportunity for influential, individual choice now.

Summary

Main symbols introduced

Symbol	Unit	Description
C_{gs}	F	gate–source capacitance
f	Hz	frequency
g_m	$A\,V^{-1}$	mutual conductance
g_o	Ω^{-1}	output conductance
$Q_n(y)$	$C\,m^{-2}$	inversion layer charge density at position y
V_s	V	surface potential at channel position y
$\bar{\mu}_e$	$m^2\,V^{-1}\,s^{-1}$	surface electron mobility

Subscripts

D, d	drain
G, g	gate
S, s	source

Main equations

$$I_{SD} = \frac{-W\bar{\mu}_e C_{ox}}{L}[(V_{GS} - V_{TH})V_{DS} - V_{DS}{}^2/2] \quad \text{(before pinch off)} \tag{15.7}$$

$$I_{DS(sat)} = \frac{-W\bar{\mu}_e C_{ox}}{2L}(V_{GS} - V_{TH})^2 \quad \text{(after pinch off)} \tag{15.9}$$

$$\epsilon_{ox}\mathscr{E}_{ox} = \epsilon_s\mathscr{E}_s \tag{15.16}$$

Exercises

1. Sketch a cross-section, top view and I_{DS}/V_{DS} characteristic for an enhancement mode p-channel MOSFET. What is the polarity of the threshold voltage?

2. An n-channel MOSFET has a threshold voltage of 1 V and applied voltages $V_{GS} = 4$ V, $V_{DS} = 7$ V. Name the device type and state whether or not it is in saturation. Find I_{DS} if $g_m = 2$ mA V^{-1} under these conditions.

3. In most real MOS integrated circuits the source and substrate are not shorted in all devices, and a reverse bias is imposed instead. What will be the effect on V_{TH} in a p-channel device? in an n-channel device? (This is known as the 'body effect'.)

4. Choose a MOSFET type and select W, L and t_{ox} given that $\bar{\mu}_e = 0.07$, $\bar{\mu}_h = 0.03$, and $V_{TH} = -1$ V if $I_{DS} = -20$ mA when $V_{DS} = -10$ V and $V_{GS} = -5$ V.

5. Find I_{DS} in a p-channel MOSFET if $W = 40$ μm, $L = 5$ μm, $\bar{\mu}_h = 0.02$ m^2 V^{-1}s^{-1}, $t_{ox} = 80$ nm, $V_{TH} = -1.5$ V, $V_{GS} = -6$ V and $V_{DS} = -3$ V.

6. Devise a two-input NAND gate using two enhancement mode n-channel MOSFETs and an enhancement mode MOS load. Sketch a layout for the doping and the metallization.

7. A MOSFET has an aspect ratio $W/L = 10$, $\bar{\mu}_e = \bar{\mu}_h = 0.2$ m^2 V^{-1}s^{-1} and $C_{ox} = 0.01$ F m^{-2}. If the magnitude of $I_{DS(sat)}$ is 62.5 mA for $V_{GS} = 1$ V and 122.5 mA for $V_{GS} = 2$ V, find the threshold voltage and the transistor type.

8. The expression for $I_{DS(sat)}$ can be used to find V_{TH} from the I_{DS}/V_{DS} characteristics. Do this for the following data, and find $\bar{\mu}_e$ if $W = 100$ μm and $L = 8$ μm.

V_{GS}	V	2	3	4	5
$I_{DS(sat)}$	mA	0.013	0.46	1.5	4.6

9. Draw a series of cross-sectional diagrams illustrating the stages in fabricating the CMOS inverter of Figure 15.10.

10. Estimate the minimum channel length and gate oxide thickness for a MOSFET with 2.10^{22} m^{-3} channel doping if $V_{GS(max)} = V_{DS(max)} = 10$ V. Use data from section 15.4.

APPENDIX

1 Crystal structure

Band structure is exhibited most perfectly by crystalline materials, in which the atoms are arranged in regular patterns. The smallest three-dimensional component of crystal which contains the full pattern is called the unit cell. The unit cell for sodium chloride is a cube with alternating sodium and chlorine atoms at the vertices (Figure A1.1). That of silicon is also a cube, but far more complex (Figure A1.2). Silicon atoms reside at the vertices, at the centres of each face, and at internal points (18 atoms in all). Each atom is at the centre of a tetrahedron, 0.235 nm from its four nearest neighbours. Diamond crystals are arranged in the same way. Gallium arsenide has a similar structure, with Ga and As atoms alternating – the zinc blende pattern.

Crystal directions and planes in a cubic lattice are specified in terms of sets of three integers known as Miller indices. Different styles of bracket distinguish between directions and planes. To derive the Miller indices of a plane:

1. Draw right-handed Cartesian (x, y, z) axes from one corner of a unit cell. Take the cell size as 1.
2. Note where the plane intersects the (x, y, z) axes. (No intersection is given as infinity.)
3. Take the reciprocals of these numbers. Should the answer be negative, it is shown by a bar over the integer.

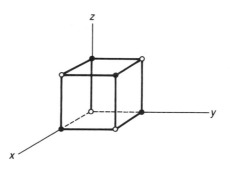

Figure A1.1 Unit cell for a cubic lattice e.g. sodium chloride

Figure (A1.3) gives examples of (100) and (111) planes. The outward normal directions to these planes are termed the [100] and [111] directions. Since the lattice may possess several symmetries, the term {100} denotes all

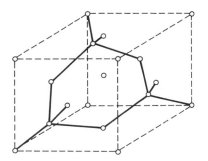

Figure A1.2 Unit cell for silicon

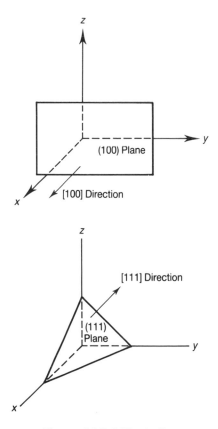

Figure A1.3 Miller indices

planes equivalent to the (100) plane, and $\langle 100 \rangle$ means all directions equivalent to [100].

In silicon the $\{111\}$ planes have the highest atom density, so $\langle 111 \rangle$ material is preferred for epitaxy and hence for bipolar devices. $\langle 100 \rangle$ material is commonly used for MOS integrated circuits, because the low surface density of atoms produces a lower interface state density.

APPENDIX

2 Acronym index

BARRITT	barrier injection and transit-time diode
BCCD	buried channel charge-coupled device
BHF	buffered hydrofluoric acid
BJT	bipolar junction transistor
BNDC	bulk negative differential conductivity
CCD	charge-coupled device
CDI	collector diffusion isolation
CMOS	complementary MOS
CRT	cathode ray tube
CVD	chemical vapour deposition
DDE	double diffused epitaxy
DIMOS	double implanted MOS
DRAM	dynamic random access memory
DSW	direct step on wafer
EAROM	electrically alterable ROM
ECL	emitter-coupled logic
EEPROM	electrically erasable programmable ROM (also E^2PROM)
EHP	electron–hole pair
EPROM	electrically programmable ROM
FAMOS	floating-gate avalanche MOS
FET	field effect transistor
HEXFET	hexagon cell FET
HMOS	high performance MOS
IC	integrated circuit
IGFET	insulating-gate FET
I^2L	integrated injection logic
IMPATT	impact avalanche transit-time device
ITO	indium tin oxide
JEDEC	joint electron device engineering council
JFET	junction FET
LASCR	light-activated silicon-controlled rectifier
LASER	light amplication by stimulated emission of radiation
LEC	liquid-encapsulated Czochralski process
LED	light-emitting diode
LPCVD	low pressure chemical vapour deposition

LPE	liquid phase epitaxy
LSI	large-scale integration
MBE	molecular beam epitaxy
MESFET	metal–semiconductor FET
MIM	metal–insulator–metal
MISFET	metal–insulator–semiconductor FET
MNOS	metal–nitride–oxide–semiconductor
MOS	metal–oxide–semiconductor
MOSFET	metal oxide semiconductor field effect transistor
MOST	MOS transistor
MTBF	mean time before failure
MTL	merge transistor logic $(= I^2L)$
NMOS	n-channel MOS
NPN	n-type, p-type, n-type semiconductor
NTD	neutron transmutation doping
PECVD	plasma-enhanced chemical vapour deposition
PIN	p-type, intrinsic, n-type semiconductor
PMOS	p-channel MOS
PN	p-type, n-type semiconductor
PNP	p-type, n-type, p-type semiconductor
PROM	programmable read only memory
PUT	programmable unijunction transistor
RAM	random access memory
RIE	reactive ion etching
RMOS	resistor load MOS
ROM	read only memory
SAMOS	self-aligned MOS or stacked-gate avalanche MOS
SBD	Schottky barrier diode
SCR	silicon- (or semiconductor-) controlled rectifier
SCS	silicon-controlled switch
SEM	scanning electron microscope
SIMS	secondary ion mass spectroscopy
SOAR	safe operation area
SOI	silicon on insulator
SOS	silicon on sapphire
SRH	Shockley Read Hall recombination statistics
TED	transferred-electron device
TFT	thin film transistor
TRAPATT	trapped plasma avalanche-triggered transit
UJT	unijunction transistor
ULSI	ultra large-scale integration
VLSI	very large-scale integration
VMOS	v-groove MOS
VPE	vapour phase epitaxy

3 Further reading

Books

Bar-Lev, A., *Semiconductors and Electronic Devices*, 2nd edn, Prentice Hall, 1984.

Grove, A. S., *Physics and Technology of Semiconductor Devices*, John Wiley, 1967. (A bit dated, but excellent on the basics of *pn* junctions.)

Millman, J., *Microelectronics: Digital and analog circuits and systems*, McGraw Hill, 1979. (This concentrates more on circuit aspects.)

Morgan, D. V. and Board, K., *An Introduction to Semiconductor Microtechnology*, John Wiley, 1983. (A good introduction to the fabrication of ICs.)

Seymour, J., *Electronic Devices and Components*, Pitman, 1981. (A good basic text with plenty of examples.)

Streetman, B. G., *Solid State Electronic Devices*, 2nd edn, Prentice Hall, 1980. (A superb text with more of a physics approach.)

Sze, S. M., *Physics of Semiconductor Devices*, 2nd edn, John Wiley, 1981. (A more advanced text, suitable for project work and postgraduate study.)

Sze, S. M., (ed.), *VLSI Technology*, McGraw Hill, 1983. (An advanced book on IC fabrication – a classic.)

Taub, H., and Schilling, D., *Digital Integrated Electronics*, McGraw Hill, 1977. (Another text which concentrates more on circuit aspects.)

Young, E.S., *Fundamentals of Semiconductor Devices*, McGraw Hill, 1978. (Priceless.)

Articles

A few original articles are listed for those who want to dig into the earlier foundations of the subject. Review articles, especially those from *Scientific American* provide a refreshing alternative to textbooks.

Bardeen J, and Brattain, W. H., 'The transistor, a semiconductor triode', *Physical Review*, vol. 74, 230 (1948).

Boyle, W. S., 'Light wave communications', *Scientific American*, vol. 237(2), 40 (1977).

Ehrenreich, H. 'The electrical properties of materials', *Scientific American*, vol. 217 (3), 195 (1967).

Sah, C. T., Noyce, R. N. and Shockley, W. 'Carrier generation and recombination in *pn* junction and *pn* junction characteristics', *Proceedings of the Institute of Radio Engineers*, vol. 45, 1288 (1957).

Shockley, W., 'The theory of *pn* junctions in semiconductors and *pn* junction transistors', *Bell Systems Technical Journal*, vol. 28, 435 (1949).

Thomas, G. A., 'An electron–hole liquid', *Scientific American*, vol. 234(6), 28 (1976).

Vacroux, A. G., 'Microcomputers', *Scientific American*, vol. 232 (5), 32 (1975).

Special issue on microelectronics: *Scientific American*, vol. 237 (3), (1977).

The Institute of Electrical and Electronic Engineers. *Journal of Solid State Circuits* has also devoted issues to special topics:

August 1977 and August 1978, *Semiconductor Materials and Processing Technology*.

April 1979, *Very Large-Scale Integration*.

October 1977, 1978 and 1979, *Semiconductor Memory and Logic*.

One word of encouragement: read all you like but *do the numerical examples*. Knowledge must be used in order to possess it – and to pass examinations.

Index